COMPUTER
SOFTWARE APPLICATIONS
IN CHEMISTRY

COMPUTER SOFTWARE APPLICATIONS IN CHEMISTRY

PETER C. JURS

Department of Chemistry
The Pennsylvania State University
University Park, Pennsylvania

A Wiley-Interscience Publication

JOHN WILEY & SONS

New York • Chichester • Brisbane • Toronto • Singapore

Library of Congress Cataloging in Publication Data:

Jurs, Peter C.
Computer software applications in chemistry.

Includes bibliographies and index.
1. Chemistry—Data processing.   2. FORTRAN (Computer
program language)   I. Title.

QD39.3.E46J873   1987   542'.8   86-11088
ISBN 0-471-84735-6

Printed in the United States of America

10 9 8 7 6 5 4 3

PREFACE

Computers have become an integral part of many aspects of modern life, including the conduct of science. Chemistry is no exception. As with other tools, computers affect chemistry in a variety of ways, some of which are obvious and some of which are hidden from casual inspection. For example, interacting with the Chemical Abstracts Service on-line information retrieval service inherently requires some computer skills, whereas running a modern Fourier transform infrared spectrometer does not, even though the instrument is fully automated. Now that computer technology has become so important a part of chemistry, it is essential that chemists have sufficient computer skills to practice their profession. To meet this need, new material is being introduced into standard chemistry courses, and new courses are being developed. Books of all types now include computer-oriented topics for use in formal course work or self-study.

The two major subdivisions of computer applications in chemistry are (a) hardware interfacing and laboratory usage, where the computer becomes part of the instrument, and (b) software usage involving numerical calculations and abstract information processing tasks. This book will focus on the second of these areas. A number of good books that deal with the laboratory uses of computers are available, and the reader is referred to them for coverage of those aspects of computers in chemistry.

The aim of this book is to provide an overview or survey of computer software applications in chemistry. The coverage of each topic will be introductory, with some advanced material added periodically to selected sections. Leading references to textbooks, review articles, and the primary scientific literature will be given throughout the book for the reader to use in finding more advanced discussions of the points raised. Many sections of the book have complete programs that perform the computation being discussed, often at a level that is useful in chemical research settings. Each such program is presented with a complete input and output sequence so that the reader could implement it.

Full use of this book as a textbook in a formal course or in a self-paced situation will require access to a minicomputer or personal computer with a FORTRAN compiler. I have chosen FORTRAN as the language to use because it is by far the most widely available, scientifically oriented language. There are certainly other languages that are well suited, possibly even better suited, to certain aspects of chemical computing (e.g., Pascal and LISP), but FORTRAN is more widely available, more compatible from one machine configuration to another, and has the largest backlog of prior software.

PETER C. JURS

University Park, Pennsylvania
August 1986

CONTENTS

List of Programs and Subroutines   xi

Definitions of Programs and Subroutines   xiii

PART I. INTRODUCTION   1

    1. Introduction / 3

        1.1. Characteristics of Digital Computers / 3
        1.2. Scientific Computer Uses / 6
        1.3. Algorithm Design / 9
        1.4. Software and Programming / 11
        1.5. Job Processing / 15

    2. Error, Statistics, and the Floating-Point Number System   19

        2.1. Error and Statistics / 19
        2.2. Floating-Point Number System / 26

PART II. NUMERICAL METHODS   31

    3. Curve Fitting / 33

        3.1. Fitting of Linear Equations / 33
        3.2. General Polynomial Equation Fitting / 39
        3.3. Linearizing Transformations / 42
        References / 52

4. Multiple Linear Regression Analysis / 53

4.1. The Basic Method / 53
4.2. Implementation of Multiple Linear Regression
Analysis / 57
References / 61

5. Numerical Integration / 62

References / 71

6. Numerical Solution of Differential Equations / 72

6.1. First-Order Differential Equations / 72
6.2. Systems of Differential Equations / 76
6.3. Predictor–Corrector Methods / 83
References / 84

**7. Matrix Methods and Systems of Linear
Equations / 85**

7.1. Methods for Solving Systems of Linear
Equations / 85
7.2 Implementation of the Gauss–Jordan
Method / 92
References / 95

**8. Random Numbers and Monte Carlo
Simulation / 96**

8.1. Pseudorandom Number Generation / 96
8.2. Monte Carlo Simulation / 98
References / 123

9. Simplex Optimization / 125

9.1. The Simplex Method / 125
9.2. Chemical Applications of Simplex Optimization / 131
9.3. Nonlinear Least-Squares Data Fitting by Simplex / 133
References / 140

PART III. NONNUMERICAL METHODS   **141**

10. Chemical Structure Information Handling / 143

10.1. Introduction / 143
10.2. Wiswesser Line Notation / 145
10.3. Connection Table Representation / 150
References / 155

11. Mathematical Graph Theory / 156

11.1. Introduction / 156
11.2. Registration of Chemical Compounds / 158
11.3. Enumeration of Isomers / 162
11.4. Atomic and Molecular Path Counts / 162
References / 169

12. Substructure Searching / 171

12.1. Principles of Substructure Searching / 171
12.2. A Working System: CAS ONLINE / 178
References / 178

13. Molecular Mechanics / 179

13.1. Introduction / 179
13.2. Implementation of the Method / 180
References / 185

14. Pattern Recognition / 186

14.1. Pattern Recognition Methods / 186
14.2. Selected Chemical Applications of Pattern
        Recognition / 207
References / 209

15. Artificial Intelligence and Expert Systems / 212

References / 217

**16. Spectroscopic Library Searching and Structure
    Elucidation / 219**

16.1. Mass Spectrometry / 219
16.2. Infrared Spectra / 223
References / 227

PART IV. GRAPHICS   **231**

17. Graphical Display of Data / 233

References / 242

18. Graphical Display of Molecules / 243

References / 249

Index / 251

LIST OF PROGRAMS AND SUBROUTINES

Chapter	Program	Subroutines
3	EXPFIT	LLS,TRANS
	MICMEN	
4	MREG	
5	NISR	
6	RK4	
	RUNKUT	PLOT
7	—	GJE
	LINEQ	DLIN
8	—	RAND
8	QMAS	
9	SIMPLX	ERROR
10	CTTEST	CTIS, CTPR
11	PATH	ALPATH
14	LM	TRAIN, PRED

DEFINITIONS OF PROGRAMS AND SUBROUTINES

EXPFIT—Program to fit an exponential function to a set of data using subroutines TRANS and LLS

LLS—Subroutine that performs a weighted linear least-squares fit to a set of supplied data

TRANS—Subroutine that performs the exponential to linear and reverse transformation for exponential function fitting

MICMEN—Program to extract Michaelis–Menten constants from sets of kinetic data by three different approaches

MREG—Program to perform multiple linear regression analysis of sets of data

NISR—Program to perform a numerical integration of a supplied equation using Simpson's rule

RK4—Program to perform a numerical solution for a differential equation using the fourth-order Runge–Kutta method

RUNKUT—Program to solve linked sets of first-order differential equations using the fourth-order Runge–Kutta method

GJE—Subroutine to solve a set of linear equations using a simplified version of Gauss–Jordan elimination

LINEQ—Program to solve a set of linear equations by calling the subroutine DLIN

DLIN—Subroutine that solves a set of linear equations by Gauss–Jordan elimination

RAND—Subroutine implementing a multiplicative method for the generation of pseudorandom numbers

QMAS—Program that simulates a quadrupole mass filter and generates ion trajectories

SIMPLX—Program that implements the simplex optimization procedure for fitting an equation to a set of data

ERROR—Subroutine used by SIMPLX for evaluation of the error function being minimized

CTTEST—Routine for input and output for testing of connection table subroutines

CTIS—Connection table input subroutine

CTPR—Connection table printing routine

PATH—Program to enumerate atomic and molecular paths by calling subroutine ALPATH

ALPATH—Subroutine that takes a connection table and generates atomic and molecular paths

LM—Program implementing the learning machine pattern recognition method by calling subroutines TRAIN and PRED

TRAIN—Subroutine that generates a discriminant from a set of data using the learning machine algorithm

PRED—Subroutine that uses a supplied discriminant to predict the class of unknown data points

COMPUTER
SOFTWARE APPLICATIONS
IN CHEMISTRY

PART I

INTRODUCTION

1

INTRODUCTION

1.1 CHARACTERISTICS OF DIGITAL COMPUTERS

Modern digital computers are uniquely capable devices. They can manipulate and transform information in ways that carry out functions that were previously performed only by the human brain. The power and breadth of application of digital computers is a result of the fact that they *simultaneously* possess a number of mutually supportive properties. The following paragraphs will discuss these individually, but it is the combination of these properties at the same time in the same machine that leads to the uniqueness of the modern digital computer.

High-Speed Operation

High-speed operation is the property most often associated with computers. While speed is important, speed alone is useless without associated properties as well. Nonetheless, very large computations can be performed routinely with fast machines. It is routine for computers to perform one million operations per second, or one operation per microsecond. It is very difficult to comprehend the quantity one million, so we must resort to analogies and comparisons, as in the following. There are as many microseconds in one hour as there have been full seconds in the 120 years that have passed since the end of the American Civil War in 1865.

Low Error Rate

The low-error-rate property refers to the fact that computers rarely (in fact, almost never) introduce errors into their operations through malfunctions. This is due primarily to high-reliability components and partly to system

3

design and software design that continually checks for such errors and announces them when they do occur. This type of error must be distinguished from errors that result from poor user software design, as will be discussed in a later section.

Varied Representation of Information

Many people consider computers as merely "fast adding machines." While computers can do arithmetic computations extremely quickly, they are in fact much more than numeric calculators. Any information whatsoever can be rendered into a form suitable for storage, manipulation, retrieval, and display by digital computers. Thus, they are universal information processors.

There are usually at least two different types of storage of numerical information in digital computers: integer (fixed point) and real (floating point). In the integer storage mode a fixed number of bits (often a multiple of 8 since computers are often designed with 8-bit bytes as the building blocks) are allocated for storage of a value. One bit is set aside for the sign, leaving the remaining bits for representing the absolute value. In a typical computer, 4 bytes, or 32 bits, are the default number assigned to integer storage, so there would be 31 bits allocated to storage of the value. Thus, numbers in the range -2^{31} to $2^{31} - 1$ can be stored. The precision of storage of a number in the integer mode depends on the absolute value of the number.

The real storage mode achieves two important goals: scaling and constant precision. Real storage mode uses a logarithmic notation much like scientific notation. In a typical computer, 4 bytes, or 32 bits, are assigned to a single-precision real value. One bit is allocated for the sign of the value, 7 bits for the exponent, and 24 bits for the fraction. In this representation, numbers between about 2.4×10^{-78} and 7.2×10^{75} can be stored, each with a constant precision of 1 part in 2^{24}. The scientific notation actually used in many computer systems is based on the octal or hexadecimal number systems. This is because numbers expressed in octal or hexadecimal are easily interconvertible to binary. Table 1 shows some numbers expressed in their binary, octal, decimal, and hexadecimal forms. A more detailed discussion of the floating-point number system is presented in a subsequent section.

General alphanumeric information can be stored as codes. Two of the best known such codes are EBCDIC (Extended Binary Coded Decimal Interchange Code) and ASCII (American Standard Code for Information Interchange). Each of these uses 2-byte, or 16-bit, patterns to represent each letter, number, and special character that must be coded. The actual code patterns can be found in many standard references. Thus, any alphanumeric information can be stored and therefore manipulated as well.

TABLE 1. Number System Bases and Conversions

Binary	Octal	Decimal	Hexadecimal
0	0	0	0
1	1	1	1
10	2	2	2
11	3	3	3
100	4	4	4
101	5	5	5
110	6	6	6
111	7	7	7
1000	10	8	8
1001	11	9	9
1010	12	10	A
1011	13	11	B
1100	14	12	C
1101	15	13	D
1110	16	14	E
1111	17	15	F
10000	20	16	10
10001	21	17	11

Stored Program Capability

Stored program capability frees the digital computer from the slow world of humans as it executes its instructions. Once the sequence of instructions to be executed is all stored internally, then the instructions can be performed at electronic speed. This is the difference between a hand-held calculator performing calculations as the user punches buttons one at a time and a digital computer executing an entire program. The idea of storing the program was an insight of John von Neumann, one of the pioneers of computation, in the 1940s. He correctly saw that once the instructions were stored, they could be operated on just as could data. This insight led von Neumann directly to the concept of conditional transfer, the next crucial property of digital computers.

Conditional Transfer

Conditional transfer means that an executing program can test the values of intermediate results, and it can then take one of several actions depending on the results of the test. This allows computer programs to break out of lock-step, the routine execution of fully determined sequences of statements. It is conditional transfer that allows computers to behave flexibly.

Digital Operation

The development of modern digital computers has been coupled with an emphasis on discrete mathematics. It has forced a rethinking of much of mathematics that heretofore was concerned with continuous functions much more than discrete mathematics.

Electronics and Technology

The basic ideas incorporated into computers existed well before they could be rendered in a practical device. Then fueled by the necessities of World War II and the availability of vacuum tubes, electronics made computer implementation possible. The developments of computer technology can be divided into five generations that are defined by their dominant technologies. The first generation was characterized by the use of vacuum tubes, the second by transistors, the third by integrated circuits, and the fourth by very large scale integrated circuits (VLSI). We are currently (1985) at the end of the third generation. The latter part of the 1980s will be the fourth generation, and the fifth generation will begin in the 1990s. The basic design of the computers of the first four generations is that of one central processing unit (CPU) connected to a main memory. This makes such computers inherently serial devices, although the rate of operations can be very fast. Such serial machines have been termed von Neumann machines. The fifth generation of computers will involve designing them without this inherent serial nature and incorporating artificial intelligence in the systems. Thus, while the first four generations of computers relied on electronic advances, the fifth generation will rely more on new ideas for the architecture of the systems. In addition to electronics, other scientific and technological areas have contributed to advances in computer design—crystallography, optics, plasma physics, polymer chemistry, semiconductor technology, surface physics and chemistry, and others.

Suggested Reading

L. M. Branscomb, "Information: The Ultimate Frontier," *Science*, **203**, 143 (1979).
R. M. Davis, "Evolution of Computers and Computing," *Science*, **195**, 1096 (1977).
H. A. Simon, "What Computers Mean for Man and Society," *Science*, **195**, 1186 (1977).

1.2 SCIENTIFIC COMPUTER USES

Digital computers are being used in nearly all phases of modern science, and chemistry is certainly no exception. The infusion of computer methodology into science has occurred by degrees during the period from 1950 to be present. Recently, computers have come to be important in nearly all phases of scientific experimentation and inquiry, and their presence in so many phases of science can tend to blur their importance. The following list of some of the most prominent uses of digital computers in chemistry shows the breadth of usage and importance of these new scientific instruments.

Numerical Calculations

The first and most obvious use of computers was in numerical calculations, and it continues to be extremely important. The availability of powerful computers capable of performing extremely large computations has altered the nature of a number of scientific fields, including weather forecasting, nuclear reactor design and control, economic modeling, control of weapons systems, manned spacecraft design and use, environmental monitoring, cryptography, telephone switching networks, and so on. Many calculations are done routinely now that were inconceivable prior to the availability of large computers.

The comparison between the rate of computation of a person and a computer is instructive. Consider a hypothetical computer that can perform one million operations per second. Multiply this by 3600 sec/hr to get 3.6×10^9 operations/hr. For a person, suppose an arithmetic computation rate of 1 operation/sec. Multiplying this by 3600 sec/hr, 8 hr/day, 5 days/ week, 50 weeks/year, and by 70 years to get 4×10^8 operations/lifetime. Thus, there is a very rough equivalence between one hour of machine time and one person's lifetime.

In chemistry, there are many areas in which the capability to perform large numbers of computations has become crucial. Examples of such fields include X-ray crystallography, quantum mechanics, neutron scattering, liquid simulations, reaction dynamics, and many others.

Time-Limited Problems

The dividing line between numerical calculations and time-limited problems is indistinct. However, some modern chemical experiments must have fast computations coupled to them in order to allow them to be attempted at all. The nature of many problems requires this.

For example, to perform Fourier transform spectroscopy [such as Fourier transform infrared spectroscopy or Fourier transform nuclear magnetic resonance (NMR) spectroscopy] requires the rapid transformation of large data arrays from the time to frequency domain. The techniques of Fourier transform spectroscopy were all known well in advance of the availability of fast digital computers, but the techniques were not generally feasible. They waited for inexpensive, fast computing and the discovery of the fast Fourier transform, a software breakthrough, before they became widely used.

Modeling, Simulation, and Optimization

Modeling and simulation of chemical experiments, instruments, and systems is now routinely done to understand the interactions between the adjustable parameters and to search for the best values for them. Computers can be used in simulations to design chemical experiments or instruments and then to optimize them with respect to the adjustable parameters. This can be markedly more economical than building prototype experiments or instruments to be used for optimization.

Information Storage and Retrieval

Chemistry is often characterized as the scientific field with the best information organization. Within the domain of bibliographic information, this is largely because of the American Chemical Society's Chemical Abstracts Service (CAS), which has systematically abstracted the worldwide chemical literature. In the endeavor, CAS has come to employ computerization to keep up with the enormous demand for such services. The type of information stored by CAS is both structural and bibliographic. Bibliographic information consists of titles, authors, journal citations, abstracts of papers and patents, and so on. The structural information stored is the molecular structures of the compounds reported in the literature abstracted. The computer software used to print *Chemical Abstracts* and many of the ACS journals must handle the structural representations used internally by the computers. Other large banks of computer-compatible structural representations and data exist both in the public sector and within industrial companies.

The second major type of information repository is spectral data banks. Here, the data stored, manipulated, retrieved, and displayed are spectra of compounds such as mass spectra, infrared spectra, or proton or ^{13}C NMR spectra. A number of commercial services offer software that allows searching of such data banks by customers who are working on structure elucidation problems.

Experiment Management and Control

Most modern chemical instrumentation contains a computer as an integral part. This feature may be advertized as an advantage or the computer may be buried within the instrument so that it is hidden from the user. It is obvious that modern instrumental chemistry has become utterly dependent on computerization. Early uses of computers in instrumentation included data-logging capabilities alone. More recently, the microcomputers or minicomputers imbedded within instruments have been endowed with a modicum of intelligence that allows some error checking, routine recalibration, adjusting for blank spectra or dark currents, and other such necessary steps in instrumental usage. This has led to the term *computer-aided chemistry* as applied to the laboratory.

Intelligent Problem Solving

Computers are certainly capable of performing prodigious amounts of computation, and this characteristic leads to important consequences in the way in which science is now conducted. However, computers are much more than fast adding machines; they are universal information processors and thus can be used for many intrinsically nonnumerical tasks. Computers can certainly be used to tackle tasks that are normally thought of as requiring human intelligence. Perhaps the primary chemical example is that of elucidation of chemical structure from spectroscopic and other data. This can be done partly by library searching of data banks, but it also requires the use of high-level inference and true intelligence. There have been several research projects tackling this area of computer usage in chemistry, and this area is certain to grow dramatically in the decades ahead.

Graphical Display of Data and Molecular Structures

When computers are provided with graphical display devices, they can provide the user with new ways of seeing the results of computations or experiments. The display of complex data in graphical form is not new, but the ability to manipulate the display, scale it, rotate it, use hidden line plots, and so on, is new. These abilities are important in allowing the experimenter to deal interactively with the results, and they make the complete cycle of theory, experiment, and theory go much faster. The ability to display molecular structures on computer-controlled displays is really new. For the first time, chemists can view complex molecular structures while they are rotated, scaled, put in contact with other structures, and so on. There are many things that physical models of structures cannot be used for that are routine in computer displays. For example, with a computer display of an enzyme and its substrate, the user can investigate the fit of one with the other as a function of the relative locations of the two molecules. If unimportant side chains in either of the molecules obscure the view, these portions of the structures could be suppressed. The chemists can tour the inside of large macromolecules such as DNA in a way never before possible.

1.3 ALGORITHM DESIGN

The most important aspect of applying computers to the solution of problems is that this task forces the scientist to define the problem completely, analyze it fully, and understand it thoroughly. The first step in attacking a problem is *problem analysis*. This involves defining what is to be done and how. It may involve breaking the original problem down into subproblems. It will involve listing all the available inputs and all the desired outputs.

An *algorithm* is a procedure consisting of a well-defined, finite set of unambiguous rules giving a sequence of operations for solving a specific type of problem. It codifies the logic involved in arriving at the solution to a problem. Development of algorithms and studies of the relative merits of alternative algorithms for the solution of a given problem is one of the major areas of computer applications in science. Unlike the field of mathematics, existence theorems are not sufficient in computing. There must exist a workable method for attacking the problem, of course, but the method must be such that it can be actually implemented and executed.

Five important features of algorithms are finiteness, definiteness, input, output, and effectiveness. An algorithm must be finite insofar as it must terminate after a finite number of steps. This is especially important in iterative algorithms that make successive approximations in an attempt to solve a problem to the required level of precision. Definiteness refers to the necessity for each and every step of the procedure to be precisely defined. It is the necessity for absolute clarity of procedure that is one of the best features of using computers to solve problems—the scientist is forced to think clearly and unambiguously. Input and output refer to the algorithm's communications with the environment in which it operates. Effectiveness means that the algorithm must be practical, that is, must be capable of implementation.

An important criterion for the degree of goodness of an algorithm is its speed of execution. Of course, speed can only be considered after it is known that the algorithm generates correct results or provides messages when it cannot do so. That is, once an algorithm is known to be robust, speed is important in comparing several different algorithms that perform the same task. It is very often the case that execution speed can be traded for memory size in algorithm design. If the programmer is willing to store more intermediate results, execution time can be decreased. In which direction this trade-off should be pushed in any given situation depends on many factors such as available memory size, fee structure of the computer being used, how time critical the application is, the relative costs of writing the algorithms to use memory or conserve it, the relative costs of making the code understandable to others, and many other factors.

Suggested Reading

B. W. Kernighan and P. J. Plauger, *Software Tools*, Addison-Wesley, Reading, MA, 1976.

G. J. Myers, *Composite/Structured Design*, Van Nostrand Reinhold, New York, 1978.

G. J. Myers, *Software Reliability. Principles and Practices*, Wiley-Interscience, New York, 1976.

M. G. Walker, *Managing Software Reliability*, North-Holland, New York, 1981.

1.4 SOFTWARE AND PROGRAMMING

The task of creating good software is challenging, can be maddening, and is ultimately very rewarding. That great care must be practiced has been recognized by virtually all who attempt to develop software. Birnbaum (1982) said:

> Programming is a craft (that is, things are done as they have been done rather than as a theory would indicate). Programming is not based on a science, and programs are not bound by natural laws in their manipulation of data. The programmer may cause practically any arbitrary sequence of logical and arithmetic manipulations to be performed on the data, so long as the results remain within the range of values acceptable to the hardware.

> The computer is probably man's intellectually richest invention, making programming the most complex craft ever practiced.

The four major subsections of program writing are (a) planning, (b) coding, (c) debugging, and (d) testing. They are interrelated to a very large extent.

Program Planning

The area of program planning or program design has received a large amount of attention recently (e.g., Myers 1978). Several theories have been developed that provide methods for developing software systems that will be superior to those developed in more traditional ways. In this context, *better* means that the resultant programs will be easier to understand, change, and maintain.

The primary building block of program structure is the module. A module is a collection of executable program statements that has the following set of characteristics:

1. It is a closed routine.
2. It can be called from any other module in the program.
3. It can be compiled independently.

A module can be described by reference to its function (what it does), by its logic (how it performs its function), and by its context (its usage). These three attributes are related but may vary with respect to degree of interdependence.

A measure of how good a program is comes from its complexity. The less complex a program, the better. Complexity can be reduced by partitioning, using hierarchical structure, and maximizing independence among system parts. Partitioning is also called modularity. Partitioning a program into modules creates a number of well-defined, documented boundaries within the program. This enhances the understandability of the program. Using a hierarchical structure allows one to focus on parts of the system in whatever degree of detail is desirable. Parts of the overall system that are unimportant can be ignored. Independence refers to the fact that the individual modules of the system should be independent of one another.

An extremely important property of programs is their readability. A common set of properties shared by most readable programs has been listed (Elshoff and Marcotty 1982) as: "The program is well commented. The logical structure of the program is constructed of single-entry single-exit flow of control units. Variable names are mnemonic and references to them are localized. The program's physical layout makes the salient features of the algorithm that is implemented stand out." Elshoff goes on to recommend how to apply a set of transformations to a program in order to enhance readability. He gives a realistic example showing the benefits that accrue as a consequence of providing a readable program.

Program Coding

The coding of a program actually takes only a fraction of the total time that goes into the overall construction of the system. Novices are prone to tackle the coding portion of the tast too early in the overall process, thereby making it necessary to go back to redesign code and correct mistakes and misconceptions. The coding part of software creation will generally go much more smoothly if the planning activity was done thoroughly first.

Adoption of some standard coding techniques eases the task of code creation and later debugging. For example, it is wise to assign statement numbers in increasing sequence so that it is easy to find references to them in other parts of the program. It is common to indent all the statements that fall within a DO loop by some standard amount so that you can see at a glance which statements are contained in the loop. A minimum of GO TO statements should be used; some authorities even go so far as to recommend that programs not use GO TO statements at all. Many programmers adopt personal conventions about numbering of FORMAT statements. Some put them all at the head of a module, others put them all at the end, others put them adjacent to the READ or WRITE statements they serve. Consistent use of one particular convention makes code much easier to generate and more readable.

Debugging of Programs

The debugging of programs falls into two areas: syntax errors and logic errors. The two types of errors are fixed using quite different approaches.

Syntax Errors. Syntax errors are mistakes in the rules of the programming language being used, here FORTRAN. Novice programmers get a lot of practice chasing down syntax errors because they are easy to make and

FORTRAN compilers are completely unforgiving of such errors. Most compilers give at least some indication of what syntax errors there are in a program being compiled. Unfortunately, these messages are not always clearly worded and can be downright cryptic. Furthermore, the syntax error messages coming from the compiler may not be complete or correct. (If they were complete and correct, then the compiler could correct the errors and proceed.) One of the most common errors in FORTRAN coding is to use unmatched parentheses, which is an error by definition. Missing statement numbers, double usage of statement numbers, and DO loop problems are examples of common syntax errors.

A famous example of a disastrous syntax error occurred in the first U.S. *Mariner* mission to Venus. In the FORTRAN program controlling the rocket, the following statement appeared:

$$DO\ 3\ I = 1.3$$

A period had been mistakenly substituted for the intended comma. The FORTRAN compiler thought that a variable named DO3I was being set equal to the value 1.3. This simple error resulted in the failure of the mission. The error was partly the fault of the programmer, but it was also partly the fault of the FORTRAN language, which allows spaces buried within the names of variables. A simple error can have large consequences.

Logic Errors. Once a program is free of syntax errors and it compiles successfully and executes, this does not guarantee that the results are correct. There may still be logic errors in the code. Several methods can be used to check for logic errors on a routine basis. One is the echo checking of all input for correctness. That is, print out the input data to make sure that it went in as desired. If the data were not input correctly, there is no possibilty of correct results. Garbage in, garbage out. Another technique for checking for logic errors is to carry through a few hand calculations if this is possible.

In scientific computations, it is imperative to check the results produced by programs extremely carefully. The input variables should be given values covering the range of acceptable values. All combinations of input variables should be tried, if feasible. One method for testing software is to write other programs that supply input values to the program being tested. Then the program being tested can be executed thousands or millions of times rather than the more limited number that would be feasible through human testing.

It is common for results of computations to be obviously wrong but the cause of the problem to be elusive. A method for narrowing down where an error may be occurring is to sprinkle diagnostic output statements through the program to provide the user with intermediate results for checking. It is often convenient to put such diagnostic output statements in programs as they are constructed so that later the program can be checked more easily.

One method for doing this is to put such diagnostic output statements under the control of variables that can be changed at execution time to allow complete output or lesser output. For example, the statement

$$\text{IF (IDEBUG.GT.0) WRITE (NOUT,222) VAR}$$

will only result in output if the value of the variable IDEBUG is positive. Statements could be placed at strategic locations throughout the program. During the initial debugging of the program, the variable IDEBUG could be given the value 1, and later, when the program is funtioning well, it could be given the value 0. However, if the statements are left in the code, then at some later time, they could be reactivated merely by changing the value of IDEBUG. This approach could be carried one step further by having output statements of this sort with different cutoff values of IDEBUG for their activation. Setting the value of IDEBUG to 0 would cause none of them to produce output, setting it to 1 would produce some output, setting it to 2 would produce more output, and so on.

Many mystifying errors in FORTRAN programs occur as a result of subscript overflows. When the subscript of a variable is out of the allowed range, then any one of a number of baffling errors can occur. The program can crash because an instruction has been replaced in memory with data. Totally wrong computational results can be obtained because the data were replaced by instructions. Some compilers provide ways to check for subscript overflow automatically, which is an extremely useful feature. However, such checking does increase execution time and thus should not be used routinely unless necessary.

Program Testing

All programs should be tested as thoroughly as possible. For simple programs this can start with hand calculations. Writing programs in modules makes testing easier because each module can be checked individually before they are assembled. It is good programming practice to test each module with values of variables outside the intended ranges in order to ensure that the modules will respond gracefully to such insults rather than aborting. The importance of testing cannot be overemphasized. The most insidious errors in computer programming are those that result in output that is plausible but wrong. This can be avoided by thorough testing.

Tuning of Code for Optimization

The desire to create code that executes quickly is enticing, and many programmers seek efficiency at almost all costs. It is easy to develop code that is so clever in seeking efficiency that it lacks other valuable virtues such as maintainability, understandability, or even correctness. However, efficiency is worth some measure of effort. The high-level methods of achieving efficiency were discussed in prior sections; here we discuss the optimization of code through code "tuning" (Bentley 1984). In code tuning, the programmer locates the parts of the code that consume the execution time and then modifies that code to improve its performance. The activity of finding which part of the code consumes the largest portion of the execution time is called profiling.

Most programs contain limited sections of code that consume the majority of the total execution time. These are the sections that are worth fixing to increase efficiency of execution. The remainder of the code, which executes rarely, should not be fixed because it is not worth the effort. Once the decision is made to tune the code, then the question remains of how to do so. Bentley (1982) has addressed this question in a book on just this subject. He breaks the methods down into classes: exploiting common cases, precomputing a function, storing precomputed results, changing the structure of the data storage, exploiting an algebraic identity, and unrolling of loops to remove loop overhead. Examples of how these approaches can be used to advantage are given. It should be pointed out that there are other goals besides execution efficiency that can be sought through code tuning, for example, using the minimum memory or minimizing the number of input–output operations.

Suggested Reading

G. Bacon, "Software," *Science*, **215**, 775–779 (1982).

J. Bentley, "Code Tuning," *Commun. A. C. M.*, **27**, 91–96 (1984).

J. Bentley, *Writing Efficient Programs*, Prentice-Hall, Englewood Cliffs, NJ, 1982.

J. S. Birnbaum, "Computers: A Survey of Trends and Limitations," *Science*, **215**, 760–765 (1982).

J. L. Elshoff and M. Marcotty, "Improving Computer Program Readability to Aid Modification," *Commun. A. C. M.*, **25**, 512–521 (1982).

B. W. Kernighan and P. J. Plaugher, *The Elements of Programming Style*, McGraw-Hill, New York, 1974.

C. B. Kreitzberg and B. Schneiderman, *The Elements of FORTRAN Style: Techniques for Effective Programming*, Harcourt Brace Jovanovich, New York, 1972.

G. J. Myers, *Composite/Structured Design*, Van Nostrand Reinhold, New York, 1978.

1.5 JOB PROCESSING

To develop working software, the programmer must use both hardware (a terminal, maybe a printer) and a number of pieces of software (editors, compilers, assemblers, loaders, etc.). This section deals with the information flow that comes between the planning and coding of an algorithm and having a functioning piece of software actually executing on a machine.

The steps that a program goes through in becoming a functioning piece of software are:

1. Compilation.
2. Assembly.
3. Editing.
4. Execution.

Each of these operations is performed by a program that has been supplied either by the manufacturer of the computer being used (IBM, DEC, Apple) or a third-party software supplier.

The first step in job processing is compilation, which is illustrated in Figure 1. The compiler is a program that takes your FORTRAN program and changes it into assembler code. It works by following the logic of the FORTRAN program line by line and reformulating it in the assembler language of the computer being used. Because it must render the program into assembler, and because assembler is specific for each machine, the compiler also must be specific for each machine. In addition, the compiler generates lists of all the variables used in your program, all the statement numbers used, and all the subroutines called. During the compilation, the compiler will usually generate a listing of the FORTRAN program unless you instruct otherwise. Additionally, the compiler will generate messages to the user, for example, diagnostics about any perceived errors. Errors at compilation time are errors in FORTRAN syntax, that is, where your program does not follow the rules of the FORTRAN language. It is not necessarily true that because a program compiles successfully, it must execute correctly. However, in the development of any FORTRAN program, it is necessary to successfully compile your program before you can proceed to the next step. As it compiles your FORTRAN program, the compiler puts the generated assembly language version of your program on a disk file to be used in the next step.

The second step in job processing is assembly, which is illustrated in Figure 2. The assembler is another program that has been supplied, usually by the computer manufacturer. It takes the assembly language version of your program and changes it into the machine language of the computer

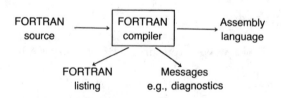

Figure 1. FORTRAN compilation.

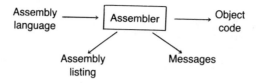

Figure 2. Assembly.

being used. The machine language is also known as object code. The object code is stored on a disk file by the assembler for further processing. In many computer systems, you can store, save, manipulate, and so on, the object code for your own programs yourself. The assembler has the ability to generate a listing of the assembly language version of your program, but most programmers never see such listings because they are very lengthy and difficult to interpret. The assembler will send any necessary messages to you such as notification of errors that have been found.

The third step in job processing is editing or link loading, which is illustrated in Figure 3. Once again, the editor or loader is a program supplied for your use by outsiders. Its job is to take the machine language version of your program and tie it together with all other necessary routines to make an executable package of machine language. The loader is notified by the prior programs (compiler and assembler) about all the subroutines that you have called in your program. These subroutines can be routines that you have written yourself and either compiled along with the main program or previously in a separate step. You may have generated a user library of your own of useful subroutines, some of which are to be used with the main program. The loader first searches your user library for needed subroutines. Then it searches the mathematics libraries or other special libraries of the computer system. Finally, it searches the libraries of FORTRAN subroutines (e.g., SQRT and input–output routines) for needed routines. If there are no unresolved calls to subroutines, the loader can put together a complete package of object code ready to be executed. The loader will put this package on a disk file for later use.

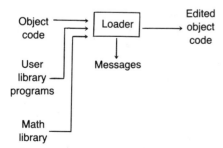

Figure 3. Editing.

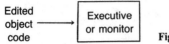

Figure 4. Execution.

    The final step in job processing is execution, which is illustrated in Figure 4. The complete object code generated by the loader is taken off disk and loaded into the memory of the computer, and the monitor turns over control of the computer to your program. Your program then begins execution and retains control of the computer system until either a STOP statement is executed or until the job is aborted by the system.

2

ERROR, STATISTICS, AND THE FLOATING-POINT NUMBER SYSTEM

2.1 ERROR AND STATISTICS

Chemistry is largely an experimental science, and as such, much of chemistry is concerned with measurements. In this book, a great many of the numerical techniques discussed deal with the handling of numbers that originated as measurements. All observed data are subject to two types of error: (1) systematic error, often called determinate error, and (2) random errors. While both types of errors must be considered in all experiments, they arise from different sources, are dealt with differently, and will be discussed separately here.

Systematic errors are those errors that result in the same error for each measurement done under exactly the same conditions. They are due to faulty instruments, calibration errors, uncompensated experimental drift, leakage in a vacuum system, parallax in reading instrumental outputs, bias of the observer, and such out-and-out errors. The amount of systematic error present in an experiment will determine the accuracy of the experiment. The *accuracy* of an experiment is how close the measurement is to the true value, the correctness of the result. The less systematic error present in an experiment, the more accurate the experiment. In principle, systematic errors can be corrected for by running blanks, using proper calibrations, and careful experimental methodology. Statistical methods generally do not deal with systematic errors.

Random error is an inherent part of the measurement process. The random errors are the small, uncontrollable fluctuations in the experimental measurements that are due to the myriad of small, uncontrollable fluctuations in the experimental conditions. That is, the independent variables of an experiment are nominally controlled, but closer examination reveals that they do fluctuate to some degree. The random uncertainties that are associated with measurements are themselves random variables. As such, they can be analyzed using statistics. The *precision* of a measurement is the reproducibility obtained when the measurement is repeated. The study of random errors using statistical methods is the purpose of this section. For much more complete discussions of these topics, see Bevington (1969) or Norris (1981).

Significant Figures

The way in which a number is written implies the precision with which the number is known. The custom is to write the number using as many digits as are significant but none beyond this. The significant digits are those that are known plus, usually, one additional digit known to be imprecise. The least ambiguous way to write numbers is with a decimal point and appropriate scaling constant, that is, in scientific notation. In this case all digits written are understood to be significant. The following numbers all have four significant figures: 6.825, 5.432×10^{-3}, 0.001327.

The concept of significant figures is introduced here because it is all too easy to have computer programs print out results with too many digits. In writing good computer software, one should be aware of the uncertainties associated with the computations being done and output results with the appropriate number of digits.

Populations, Mean, and Standard Deviation

The statistical methods for dealing with random error and characterizing it are derived from probability theory. Therefore, before coming to the real points of this section, we must take a slight detour to develop some of the underpinnings for what will follow.

When a measurement is repeated a number of times, the values obtained will usually differ somewhat. When a substantial number of measurements are taken, a distribution of values emerges. Call the individual measurement x_i. Some values for x_i are more likely than others, so a population of x_i values will be bunched about a centroid. Of the other values observed, some will be too large and some will be too small, but there will be relatively few values far from the centroid. If the measurements were repeated an infinite number of times, the parent distribution would emerge. In an actual experiment, the values observed in the repeated trials were randomly selected values drawn from the hypothetical parent distribution.

In any real experiment, a finite number of measurements are taken, and this represents a sample from the parent population. The objective of the experiment is to use the experimental data to describe the parent distribution as well as is possible. The parent population has a set of characteristics such as the mean value, median value, and standard deviation. Let μ stand for the mean value of the parent population and σ represent its standard deviation. A generalized population plotted as a probability density function

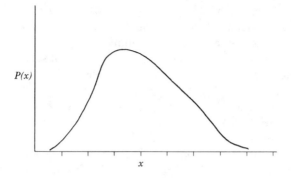

Figure 1. Probability density function.

is depicted in Figure 1. It shows the probability of observing a particular value of x as a function of x. The mean of the distribution is defined as

$$\mu = \lim_{n \to \infty} \frac{1}{n} \sum_i x_i \tag{1}$$

The width of the population is characterized by the variance or standard deviation σ:

$$\sigma^2 = \lim_{n \to \infty} \frac{1}{n} \sum_i (x_i - \mu)^2 \tag{2}$$

Thus, the standard deviation of a distribution is the root-mean-square of the deviations of the obsevations from the mean. The two parameters μ and σ are characteristic of the parent population, and their definitions depend on having an infinite number of observations.

In real experiments, one makes a finite number of measurements and uses the data to estimate the mean and standard deviation. If one makes n measurements of the quantity x, the mean is defined as

$$\bar{x} = \frac{1}{n} \sum x_i \tag{3}$$

The mean value obtained from a finite set of measurements is an estimate of the mean of the underlying parent distribution: $\bar{x} = \mu$.

The sample standard deviation s of the set of n observations of the quantity x is defined as

$$s^2 = \frac{1}{n-1} \sum (x_i - \bar{x})^2 \tag{4}$$

The denominator of the equation is $n - 1$ because this expresses the number of degrees of freedom of the equation. The number of degrees of freedom is the number of observations in excess of those needed to determine the parameters of an equation. In this equation, it takes $n - 1$ observations to determine all the x_i values plus s. Put another way, $n - 1$ is the number of degrees of freedom left after determining s from n observations.

The sample standard deviation is easily calculated using Equation (4) as shown here. It does require a two-pass algorithm, however. The first pass through the x_i accumulates the sum of the x_i values in order to calculate the mean, and the second pass accumulates the sum of the deviations from the mean in order to calculate the standard deviation.

Equation (4) can be manipulated algebraically to generate some equivalent forms that often appear in textbooks and are sometimes convenient for calculations. If the squared term in Equation (4) is expanded, this yields

$$s^2 = \frac{\Sigma\,(x_i^2 - 2x_i\bar{x} - \bar{x}^2)}{n - 1} \tag{5}$$

which can be simplified to

$$s^2 = \frac{\Sigma\,(x_i^2 - 2n\bar{x}^2 + n\bar{x}^2)}{n - 1} \tag{6}$$

and to

$$s^2 = \frac{\Sigma\,x_i^2 - (\Sigma\,x_i)^2/n}{n - 1} \tag{7}$$

and also to

$$s^2 = \frac{\Sigma\,x_i^2 - n\bar{x}^2}{n - 1} \tag{8}$$

This final expression, Equation (8), can be used to construct an algorithm to calculate both the mean and standard deviation of a set of data with one pass through the data. Such an algorithm would accumulate the sum of x_i values for calculating the mean and also the sum of squares—that is, x_i^2 values—for use in Equation (8). However, the two terms in the numerator of Equation (8) can both be large and nearly equal, which can lead to the introduction of serious errors in the computation. A discussion of the problem of loss of significance when subtracting two nearly equal numbers is given later in this chapter. In general, it is best to use Equation (4) for computation of the standard deviation. Wanek and co-workers (1982) and Solberg (1983) present some examples and discussion of this point.

Propagation of Errors

In performing numerical calculations on measured quantities, we often must plug observed values, which have associated uncertainties, into equations to obtain a computed result. Here we will discuss the proper way to propagate the uncertainties in the variables through such calculations.

In a typical case we will have a set of measured variable values, call them v_i, $i = 1, 2, \ldots, n$, and their associated standard deviations σ_i, where σ_1 is the standard deviation associated with variable v_1. We wish to compute a value for a derived parameter, y, using a function at hand:

$$y = f(v_1, v_2, \ldots, v_n) \tag{9}$$

If the variables v_i are uncorrelated, the standard deviation of y is related to the standard deviations of the individual variables by the equation

$$\sigma_y^2 = \sigma_1^2 \left(\frac{\partial y}{\partial v_1} \right)^2 + \sigma_2^2 \left(\frac{\partial y}{\partial v_2} \right)^2 + \cdots + \sigma_n^2 \left(\frac{\partial y}{\partial v_n} \right)^2 \tag{10}$$

The terms in parentheses are partial derivatives of the function with respect to each of the variables. More complete discussions of this equation can be found elsewhere (e.g., Bevington 1969).

Error Propagation Example

The molar refraction of a liquid, R, is related to its molecular weight M, refractive index n, and density d by the equation

$$R = \frac{n^2 - 1}{n^2 + 2} \frac{M}{d} \tag{11}$$

Say that experiments with benzene yielded the following values and associated uncertainties:

$$d = 0.8737 \pm 0.0002 \ g/ml$$

$$n = 1.4979 \pm 0.003$$

$$M = 78.114 \ g/mole$$

Inserting these values into Equation (11) produces $R = 26.20 \ ml/mole$.

To calculate the error in the calculated value for R, we must have the partial derivatives of R with respect to d and n. They are

$$\frac{\partial R}{\partial d} = -\frac{R}{d} = 29.98 \tag{12}$$

and

$$\frac{\partial R}{\partial n} = R \frac{6n}{(n^2 - 1)(n^2 + 2)} = 44.61 \tag{13}$$

The uncertainty in R is given by

$$\sigma_R^2 = \left(\frac{\partial R}{\partial d}\right)^2 \sigma_d^2 + \left(\frac{\partial R}{\partial n}\right)^2 \sigma_n^2 \tag{14}$$

$$\sigma_R^2 = (29.98)^2(2 \times 10^{-4})^2 + (44.61)^2(3 \times 10^{-4})^2$$
$$= 2.15 \times 10^{-4} \tag{15}$$

and

$$\sigma_R = 0.014$$

So the proper way to report the result is

$$R = 26.20 \pm 0.014$$

Equation (10) shows how error is propagated in the general case. We can derive simplified expressions for a few simple arithmetic cases as follows.

Addition and Subtraction. For the case where

$$y = av_1 + bv_2 \tag{16}$$

the partial derivatives are given by

$$\frac{\partial y}{\partial v_1} = a \qquad \left(\frac{\partial y}{\partial v_2}\right) = b \tag{17}$$

so the error in the computed result y is related to the errors in the two variables by the expression

$$\sigma_y^2 = a^2 v_1^2 + b^2 v_2^2 \tag{18}$$

This expression holds true if the two variables are not correlated. Equation (18) also describes error propagation in subtraction. Expressed in words, it says that the error in the sum of numbers is the square root of the weighted sum of squares of the errors in the variables being summed. We are combining *absolute* standard deviations.

Multiplication and Division. For the case where

$$y = av_1 v_2 \tag{19}$$

the partial derivatives are

$$\frac{\partial y}{\partial v_1} = av_2 \qquad \frac{\partial y}{\partial v_2} = av_1 \tag{20}$$

so the error in y is related to the errors in the variables by

$$\sigma_y^2 = a^2 v_2^2 \sigma_1^2 + a^2 v_1^2 \sigma_2^2 \tag{21}$$

$$\frac{\sigma_y^2}{y^2} = \frac{\sigma_1^2}{v_1^2} + \frac{\sigma_2^2}{v_2^2} \tag{22}$$

This expression is correct for division as well, and it assumes that the variables are not correlated. The expression says that for multiplication and division we combine *relative* standard deviations.

Simplified expressions such as Equations (18) and (22) can be derived easily for other simple functions such as powers, exponentials, logarithmic functions, trigonometric functions, and so on. In each case, one uses the general equation (10).

A special case of interest is the way in which error carries through an averaging calculation. Express the averaging calculation in these symbols:

$$y = \frac{1}{n} (x_1 + x_2 + \cdots + x_n) \tag{23}$$

The partial derivative of y with respect to any one of the x is given by $1/n$. Therefore,

$$\sigma_y^2 = \left(\frac{1}{n}\right)^2 \sigma_1^2 + \left(\frac{1}{n}\right)^2 \sigma_2^2 + \cdots + \left(\frac{1}{n}\right)^2 \sigma_n^2$$

$$= \left(\frac{1}{n}\right)^2 n\sigma_x^2 = \frac{\sigma_x^2}{n} \tag{24}$$

or

$$\sigma_{\bar{x}} = \frac{\sigma_x}{\sqrt{n}} \tag{25}$$

The error of the mean of n replicated measurements of a quantity is decreased by the factor of the square root of n. It takes four replicates to decreases the uncertainty by a factor of 2.

Numerical Evaluation of Error Propagation

Cases can arise when you want to estimate the way in which uncertainties are propagated through a computation but the previously described methods are not applicable. This would be the case, for example, if derivatives could not be derived. Then the following procedure might provide a reasonable estimate in the absence of more exact alternatives.

Let the function being numerically evaluated be represented by

$$y = g(x_1, x_2, \ldots, x_n) \tag{26}$$

where x is the variable and its uncertainty is represented by Δx. First calculate y using the best value for each x. Then, for all $i = 1, 2, \ldots, n$, calculate

$$y_i = g(x_1, x_2, \ldots, x_i + \Delta x_i, \ldots, x_n) \tag{27}$$

Then, set $a_i = +1$ if $y_i > y$ or $a_i = -1$ if $y_i < y$, and calculate

$$y + \Delta y = g(x_1 + a_1 \Delta x_1, x_2 + a_2 \Delta x_2, \ldots, x_n + a_n \Delta x_n) \tag{28}$$

and this final calculated value $(y + \Delta y)$ is a reasonable estimate of the contribution of propagated error. For this estimate to be good, it is necessary that the function being evaluted be continuous at the point $(x_1, x_2, \ldots, x_n)$ and that it be differentiable at that point and that the errors $\Delta x_1, \Delta x_2, \ldots$ are small.

2.2 FLOATING-POINT NUMBER SYSTEM

The real number field of mathematics is approximated in digital computers by the floating-point number system (Knuth 1969, Knoble 1979). It involves the storage of numbers in a format much like ordinary scientific notation. That is, a number is stored in two parts: one contains the significant digits, and the other contains the exponent properly scaled.

The following notation will be used to discuss the properties of the floating-point number system:

$$x = fB^{e-q} \tag{29}$$

where x is a floating-point number being stored, B is the number base of the computer, d is the computer precision, f is a base B d-digit fraction, e is a b-bit base B computer exponent and $e - q$ is the algebraic base B value of the exponent, q is the base B "excess" or exponent bias, and b is the exponent precision.

Ordinary scientific notation employs the following values for the parameters: $B = 10$ and $q = 0$. Values for these parameters can be found in references for the most popular computers (e.g., Aird 1978). For many machines, the base B is 16 (hexadecimal), $d = 6$ for single precision or $d = 14$ for double precision, and $b = 7$ bits.

For IBM computers, $B = 16$ (hexadecimal), $d = 6$ in single precision, so six base 16 mantissa digits are stored in single precision. Therefore,

$$-2^6 < e - q < 2^6 - 1$$

$$-64 < e - q < 63$$

so the number $x = 33$ would be stored in this notation as

$$x = 0.21 \times 16^{42-40}$$

because

$$x = 0.21 \times 16^2$$

$$= \left(\frac{2}{16}\right) 16^2 + \left(\frac{1}{16}\right)^2 16^2$$

$$= \left(\frac{2}{16}\right) 256 + \frac{16^2}{16^2}$$

$$= 32 + 1 = 33$$

As floating-point arithmetic is performed during the execution of a program, the values of the fraction f and the exponent $e - q$ are adjusted according to the usual algebraic laws. However, these adjustments are bounded by the precision limitations of the computer environment. Thus, there are inherent limitations to arithmetic performed using the floating-point number system.

The laws of algebra usually break down in floating-point number systems. The following is a list of the properties of the floating-point number system that must be considered to understand why this is so.

1. There is a finite set of values that a floating-point number can assume. This is because some numbers of the real number field cannot be represented in a finite number of bits. Just as one dollar cannot be exactly evenly divided into thirds in U.S. coins, many numbers cannot be represented in the floating-point number system. For example, the number $\frac{1}{10}$ is represented in binary notation by $0.0001100110011001100\ldots$ or in hexadecimal notation by $0.19999999\ldots$. In addition, several representations within the floating-point number system have the same actual value, differing only by offsetting differences in the fraction and exponents.

2. There are numbers in the floating-point number system that do not have inverses. There exist x values for which no y exists such that x multiplied by y equals 1. This property follows immediately from property 1.

3. The ordinary arithmetic operations (addition, subtraction, multiplication, division) are not necessarily closed within the floating-point number system. There are numbers x and y in a floating-point number system for which the numbers $x + y$, $x - y$, $x * y$, and x/y do not belong to the system. Just as it is not possible to evenly divide one dollar into thirds, neither is it possible to repay a one-dollar debt in three equal installments.

4. The associative and distributive laws of algebra do not necessarily hold in floating-point number systems. Thus, different results can be obtained as

a function of the sequence in which arithmetic operations are carried out on a set of variables.

These laws of algebra break down because of the loss of significant digits during arithmetic operations performed on floating-point numbers. There are at least three ways in which this can happen, any one of which can cause unforeseen errors in computations.

(a) When two numbers are multiplied, the number of significant digits needed to store the product is the sum of the number of significant digits in the two operands. However, in the floating-point number system, the product will always be stored using exactly the same number of significant figures as the operands. Thus, the result is inexact.

(b) When two numbers are added (or subtracted) in the floating-point number system, they must be scaled so that the two exponents agree. If the numbers being added have sufficiently different magnitudes, the scaling operation will cause significance loss, and $x + y = x$ for many (relatively small) values of y.

(c) When two numbers that are nearly equal are subtracted, significance loss will occur. This is because the most significant figures of the operands cancel during subtraction, and the result contains only the least significant figures.

5. The fundamental equalities and inequalities fail to hold in floating-point number systems. Operators such as the FORTRAN .EQ. and .NE. almost never agree with algebra, even after only as few as one floating-point computation.

A quantity useful in defining ways to deal with these limitations is the *machine epsilon* EPS. EPS is the smallest floating-point number for a given computer environment such that

$$1 + EPS > 1 \qquad 1 - EPS < 1$$

It is also called the *machine tolerance*.

In order to circumvent pitfalls due to the breakdown of the laws of algebra in floating-point number systems, a number of recommendations have been advanced (Knoble 1979).

Recommendation 1. Use library programs that are as robust as possible. A "robust" computer program is one that yields correct output when provided correct input or returns an error message or warning indication if it cannot. An example of a set of well-designed, general-purpose routines is the International Mathematical Statistical Library (1979).

Recommendation 2. Thoroughly test programs over the domain of possible values for input data, especially on the limits of each domain. Modules within large programs should be tested individually.

Recommendation 3. Use double-precision arithmetic unless a program's numerical computations can be proven to be stable in single precision.

Recommendation 4. For a given level of computer precision, do not input or output more decimal digits than are meaningful.

Recommendation 5. Use relative comparisons in place of absolute comparisons. Never use the FORTRAN relational operators .EQ. or .NE. for comparing floating-point quantities. Tolerant comparisons can be constructed as follows:

$$TEQ(U,V) = DABS(X-Y).LE.DMAX1(DABS(X),DABS(Y))*EPS3$$

where EPS3 is the number 3*EPS; it allows the last two mantissa bits of any two floating-point numbers x and y to be ignored during comparisons. This is because 3 base 10 = 11 base 2. Function TEQ means "If the absolute difference between double-precision arguments corresponding to parameters X and Y, when scaled relative to the unit interval, is less than EPS adjusted to ignore the two low-order mantissa bits, then the numbers are computationally equal." Thus, TEQ(X,Y) tests for computational equivalence, which is presumably the intention of the program in any case. TEQ is a logical function, and the variables X,Y, and EPS3 should all be declared as double-precision variables in the program. Tolerant comparisons can be similarly constructed for .NE., .GT., and the other relational operators.

Recommendation 6. Develop numerical algorithms that are computationally robust. That is, write programs that either avoid significance loss or else monitor it. Knoble (1979) gives an example of a FORTRAN program that incorporates robust addition and subtraction routines.

Suggested Reading

T. J. Aird, "The IMSL Fortran Converter. An Approach to Solving Portability Problems," *Lecture Notes in Computer Science*, Springer-Verlag, New York, 1978.

P. R. Bevington, *Data Reduction and Error Analysis for the Physical Sciences*, McGraw-Hill, New York, 1969.

IMSL, *International Mathematical Statistical Library Reference Manual*, IMSL, Houston, TX, 1979.

H. D. Knoble, *A Practical Look at Computer Arithmetic*, Computation Center, The Pennsylvania State University, August 1979.

D. E. Knuth, *The Art of Computer Programming*, Vol. 2, Addison-Wesley, Reading, MA, 1969, Sections 4.2 and 4.3.

A. C. Norris, *Computational Chemistry, An Introduction to Numerical Methods*, Wiley, New York, 1981.

H. E. Solberg, "Inaccuracies in Computer Calculation of Standard Deviation," *Anal. Chem.*, **55**, 1611 (1983).

P. M. Wanek, et al., "Inaccuracies in the Calculation of Standard Deviation with Electronic Calculators," *Anal. Chem.*, **54**, 1877–1878 (1982).

PART II

NUMERICAL METHODS

The paramount capability of digital computers is the very fast performance of arithmetic. Such arithmetic operations can be combined into intricate algorithms to accomplish complex tasks. When this ability to do mathematical manipulations quickly is combined with the other characteristics of computers, then the variety of tasks that can be performed is virtually limitless.

The design of algorithms was discussed in Chapter 1. In this part, we will introduce a number of specific numerical methods. Here, the computer is performing the task of carrying out computations that are arithmetic.

There is a vast difference between mathematics in general and the subfield of numerical analysis. In numerical analysis, we insist on having methods that are practical and achieve the result desired. Existence of a proof that our answer exists is not sufficient. We want to obtain the answer.

Computer algorithms solve numerical problems by applying arithmetic methods to them. Computer software cannot calculate the sine or tangent of an angle directly. Rather, it must be done by evaluating a truncated infinite series to the level of accuracy desired. The evaluation of such series is a specific example of an *iterative* method. Iterative methods abound in computer software, and we therefore describe the fundamental characteristic shared by all iterative methods. Examples of iterative algorithms will fill this chapter.

The features shared by iterative methods are as follows:

1. The means to make a satisfactory first approximation, a starting guess.
2. The means to systematically improve on the previous approximation.
3. A criterion for stopping the iterative process when sufficient accuracy has been obtained.

These features will appear many times in the algorithms that are discussed in this part.

3

CURVE FITTING

The fitting of mathematical equations to data is a routine task in experimental science. Two purposes to be served by doing this are (1) to summarize or generalize a set of data by some appropriate analytical function with a few adjustable parameters or (2) to fit fundamental, theoretically derived model equations to observed data to test hypotheses. The first purpose is empirical, and the goal is to represent the data economically. The fit equation can be used to obtain areas by integration, it can be used to obtain values for the y variable between the originally observed x values through interpolation, it can be used to obtain slopes through differentiation, or it can be used for calibration. In general, no physical significance is accorded to the fit equation in such empirical curve fitting.

In the second case, however, more care is usually invested in the derivation, and more interpretation of the results is often warranted. A physical model is being investigated, so the values obtained for the parameters are related to the physical world. Thus, one is interested in the values obtained, and one is also interested in the uncertainties of those values.

3.1 FITTING OF LINEAR EQUATIONS

The Simple Linear Case

Of the many statistical curve-fitting methods that are available, the oldest and most widely used is the method of least squares. This method depends on minimizing the sum of the squared deviations between the calculated and observed dependent variables. That is, the quantity Q is minimized:

$$Q = \sum_{i=1}^{n} (y_i - \hat{y}_i)^2 \tag{1}$$

where n is the number of data points being fitted, y_i represents the values of the observed dependent variables, and $\hat{y}_i$ represents the values of the dependent variables calculated from the fit equation:

$$\hat{y}_i = f(x_i) \tag{2}$$

Thus, we can rewrite Equation (1) as

$$Q = \sum_{i=1}^{n} [y_i - f(x_i)]^2 \tag{3}$$

The quantity Q is the sum-squared error and is often given the symbol SSE. These equations are based on the assumption that the x_i values are known exactly and the error in the observations lies in the y_i values.

The quantities involved in Equations (1)–(3) are shown schematically in Figure 1. The vertical distance between the observed point y_i and the calculated value for that point, $\hat{y}_i$, is called the deviation or error for that point. The SSE is the sum of the squares of these deviations for all the points in the data set. The least-squares fit curve is placed so that the SSE is minimized. In Figure 1, the least-squares fit curve is a straight line.

If the functional form of the equation is linear, then

$$\hat{y}_i = f(x_i) = a_0 + a_1 x_i \tag{4}$$

and

$$Q = \sum_{i=1}^{n} (y_i - a_0 - a_1 x_i)^2 \tag{5}$$

which shows explicitly that the sum-squared error Q is a function of the intercept a_0 and the slope a_1 of the fit straight line. The value of the SSE is

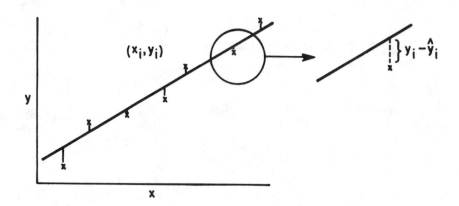

Figure 1. Plot of a data set and fitted line.

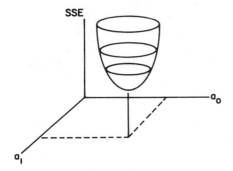

Figure 2. The sum-squared error (SSE) as a function of the model parameters a_0 and a_1.

defined for values of a_0 and a_1 over a wide range; a plot of the SSE versus a_0 and a_1 is shown in Figure 2. The SSE is a parabolic surface with a minimum at one particular pair of values for a_0 and a_1. These values, giving the minimum value for the SSE, are the values calculated by the least-squares method.

To find the values for a_0 and a_1 that minimize the SSE, we use the standard method of differential calculus—we find the partial differentials of the SSE with respect to a_0 and a_1, set these equal to zero, and solve the resulting simultaneous equations for a_0 and a_1.

The first step in the sequence involves taking the partial differential of the SSE with respect to a_0;

$$\frac{\partial Q}{\partial a_0} = 2 \sum (y_i - a_0 - a_1 x_i)(-1) = 0 \tag{6}$$

Divide through by -2 to get

$$\frac{\partial Q}{\partial a_0} = \sum (y_i - a_0 - a_1 x_i) = 0 \tag{7}$$

Break up the equation's terms to get

$$\frac{\partial Q}{\partial a_0} = \sum y_i - \sum a_0 - \sum a_1 x_i = 0$$

$$= \sum y_i - n a_0 - a_1 \sum x_i = 0 \tag{8}$$

Divide through by n to get

$$\frac{\sum y_i}{n} - a_0 - a_1 \frac{\sum x_i}{n} = 0 \tag{9}$$

or

$$a_0 = \bar{y} - a_1 \bar{x} \tag{10}$$

where $\bar{y}$ is the mean of the observed y_i values and $\bar{x}$ is the mean of the x_i values. Given a value for the slope, a_1, Equation (10) can be used to calculate the least-squares best estimate for the intercept.

Second, we must work with the intercept a_1. Take the partial differential of the SSE with respect to a_1 to get

$$\frac{\partial Q}{\partial a_1} = 2 \sum (y_i - a_0 - a_1 x_i)(-x_i) = 0 \tag{11}$$

and divide through by -2 to get

$$\frac{\partial Q}{\partial a_1} = \sum (y_i - a_0 - a_1 x_i) x_i = 0 \tag{12}$$

Multiply through and break up the equation to get

$$\sum x_i y_i - a_0 \sum x_i - a_1 \sum x_1^2 = 0 \tag{13}$$

In Equations (9) and (13), we have two equations in two unknowns, so they can be solved simultaneously for a_0 and a_1. After some algebraic manipulation, the desired result is

$$a_1 = \frac{n \sum x_i y_i - \sum x_i \sum y_i}{n \sum x_1^2 - (\sum x_i)^2} \tag{14}$$

This is the usual equation for calculating the slope. With some rearranging of the variables, another equation for a_1 can be derived:

$$a_1 = \frac{\sum (x_i - \bar{x})(y_i - \bar{y})}{\sum (x_i - \bar{x})^2} \tag{15}$$

Equations (10) and (14) or (10) and (15) can be used to calculate the slope and intrcept of the least-squares best-fit line to a set of data. The curve that is derived is unique once the set of data points is specified. That is, given the x_i and y_i values, the values of a_1 and a_0 are determined.

The equations developed to this point all include the assumption that each y_i value is as accurate as any other. That is, in Equations (1), (3), and (5) each y_i contributes an equal amount to the summation of the SSE. However, it is common in dealing with real data that some other scheme of weighting of the SSE summation is more in keeping with the characteristics of the data set. Thus, we must consider *weighted* linear least squares.

The Weighted Linear Case

For the weighted least squares, the equations are somewhat more complex than those already introduced. The expression for the SSE becomes

$$Q = \sum_{i=1}^{n} w_i (y_i - \hat{y}_i)^2 \tag{16}$$

and Equation (5) becomes

$$Q = \sum_{i=1}^{n} w_i (y_i - a_0 - a_1 x_i)^2 \tag{17}$$

The w_i are called weighting factors, and they stand in a one-to-one correspondence with the data points (x_i, y_i) themselves. The more accurately a data point is known, the larger the value for the associated w_i should be. The fit curve should pass closer to more accurately known points than to less accurately known points, and this is what inclusion of the weighting factors in Equation (16) forces to happen. The term in the summation for an accurately known point is multiplied by a larger weighting factor while the term in the summation for a less accurately known point is multiplied by a smaller weighting factor.

The weighting factors can be given values in a number of different ways, depending on the characteristics of the data set. If $w_i = 1$ for all i, then Equation (17) reduces to Equation (5); these are called absolute weights. Another option is to use relative weighting, where $w_i = 1/y_i$. Here, we are saying that the points with smaller values of y_i are known relatively accurately, and the points with larger values of y_i are less well known. A third, and very often used, weighting scheme is approriate where the uncertainties in the y_i values can be characterized by real standard deviations. In this case,

$$w_i = \frac{1}{\sigma_i^2} \tag{18}$$

where σ_i is the standard deviation of the y_i value.

Equations can be derived using the same methods employed above that express the best values for a_0 and a_1 in terms of weighted summations of the data:

$$a_0 = \frac{\Sigma w_i x_i^2 \, \Sigma w_i y_i - \Sigma w_i x_i \, \Sigma w_i x_i y_i}{\Sigma w_i \, \Sigma w_i x_i^2 - (\Sigma w_i x_i)^2} \tag{19}$$

$$a_1 = \frac{\Sigma w_i \, \Sigma w_i x_i y_i - \Sigma w_i x_i \, \Sigma w_i y_i}{\Sigma w_i \, \Sigma w_i x_i^2 - (\Sigma w_i x_i)^2} \tag{20}$$

One measure of the overall quality of the fit function is the standard deviation of the fit:

$$s^2 = \frac{Q}{n-2} \tag{21}$$

where Q is the weighted SSE of Equation (16). The denominator is the number of degrees of freedom left after fitting n data points with the two parameters of a linear equation. In general, the number of degrees of freedom of a sum of squares is the number of independent pieces of information that involve the n independent numbers $(y_1, y_2, \ldots, y_n)$ that are necessary to compute the sum of squares. The standard deviations of the intercept and slope are given by

$$s_{a_0}^2 = \frac{s^2}{D} \sum w_i x_i^2 \qquad (22)$$

$$s_{a_1}^2 = \frac{s^2}{D} \sum w_i \qquad (23)$$

$$D = \sum w_i \sum w_i x_i^2 - \left(\sum w_i x_i \right)^2 \qquad (24)$$

Another measure of the value of the fit equation is the linear correlation coefficient, which is defined as

$$R = \frac{\sum w_i \sum w_i x_i y_i - \sum w_i x_i \sum w_i y_i}{[\sum w_i \sum w_i x_i^2 - (\sum w_i x_i)^2]^{1/2} [\sum w_i \sum w_i y_i^2 - (\sum w_i y_i)^2]^{1/2}} \qquad (25)$$

The value of R ranges between -1 and $+1$. A perfect relationship between x and y yields an R of $+1$ or -1, and if there is no linear relationship between x and y, then an R value close to zero will be found. The square of the correlation coefficient is the fraction of the explained variance by the fit equation.

A number of additional indices of the quality of the fit to the data can be calculated from s^2—the standard deviation of the fit expressed by Equation (21)–and the uncertainties in the data points: σ_i. Two such indices are the chi-squared test and the F-test. They provide quantitative estimates of the quality of the fit. Detailed discussions of their derivations and interpretation can be found in the literature (Bevington 1969, Draper and Smith 1981).

Analysis of Residuals

A very effective way to detect deficiencies in fitted equations is to analyze the residuals. The residual for the ith observation is defined to be

$$e_i = y_i - \hat{y}_i \qquad (26)$$

where y_i is the observed value of the dependent variable and $\hat{y}_i$ is the value predicted from the fit equation. The residuals can be standardized by dividing them by the standard deviation of the fit

$$e_{is} = \frac{e_i}{s} \qquad (27)$$

The standardized residuals have zero mean and unit standard deviations.

The residuals should be distributed according to the normal distribution. That is, the standardized residuals should have random values distributed about zero with 95% of the values falling between -2 and $+2$. If the model that has been generated is wrong, then a plot of the residuals will often display the fact plainly. Thus, making plots of residuals and analyzing them is a powerful tool for determining whether models are good representations of your data.

Such plots can be constructed in many ways, but several popular variations are as follows: (1) plot e_{is} versus the model prediction $\hat{y}_i$; (2) plot e_{is} versus the independent variable x_i; and (3) plot e_{is} versus another possibly relevant experimental variable such as time. When plots such as these are constructed, they should show no patterns among the residuals. They should appear scattered and random if the model is a good representation of the data. Examination of residuals is one of the most generally useful tools of analysis in regression studies.

EXAMPLE OF A LINEAR FIT

A set of 10 points is to be fit with a linear equation. The points are $(0.4, 1.217)$, $(0.8, 1.821)$, $(1.25, 2.715)$, $(1.6, 2.835)$, $(2.0, 3.544)$, $(2.5, 4.778)$, $(3.1, 5.721)$, $(3.5, 6.179)$, $(4.0, 7.410)$, $(4.4, 7.067)$. The summations necesary to calculate the intercept and slope are

$$\sum x_i = 23.55 \qquad \sum y_i = 43.29$$

$$\sum x_i^2 = 72.39 \qquad \sum x_i y_i = 129.00$$

Then the intercept and slope and their deviations are

$$a_0 = 0.56 \pm 0.20 \qquad a_1 = 1.60 \pm 0.07$$

The value of the SSE is 0.30, and the value of the linear correlation coefficient is $R = 0.992$. The fit equation is

$$y = 0.56 + 1.60x$$

3.2 GENERAL POLYNOMIAL EQUATION FITTING

To this point, we have dealt with simplest linear equation as the function to be fit. The same approach can be used to fit polynomials of higher orders to data. The general polynomial of order m is

$$f(x_i) = \sum_{j=0}^{m} a_j x_i^j \tag{28}$$

Substituting this function form into Equation (16) yields

$$Q = \sum_i w_i \left[y_i - \sum_j a_j x_i^j \right]^2 \tag{29}$$

The objective is to find the values for the a_j that minimize Q. To minimize, we differentiate with respect to each adjustable parameter and set it equal to zero to get a set of normal equations,

$$\frac{\partial Q}{\partial a_k} = 0 = \sum_i w_i \left[y_i - \sum_j a_j x_i^j \right] x_i^k \qquad k = 0, 1, 2, \ldots, m \tag{30}$$

We have a set of $m + 1$ equations in $m + 1$ unknowns, so they can be solved. The equations can be most easily expressed in matrix form, so this way of defining the problem will now be developed.

As a specific case, we take the polynomial equation of degree 2, the quadratic:

$$f(x_i) = b_0 + b_1 x_i + b_2 x_i^2 \tag{31}$$

In terms of this function, the normal equations become

$$\sum_i w_i (y_i - b_0 - b_1 x_i - b_2 x_i^2) x_i^k = 0 \qquad k = 0, 1, 2 \tag{32}$$

Expanded, these become

$$b_0 \sum w_i + b_1 \sum w_i x_i + b_2 \sum w_i x_i^2 = \sum_i w_i y_i$$

$$b_0 \sum w_i x_i + b_1 \sum w_i x_i^2 + b_2 \sum w_i x_i^3 = \sum w_i x_i y_i \tag{33}$$

$$b_0 \sum w_i x_i^2 + b_1 \sum w_i x_i^3 + b_2 \sum w_i x_i^4 = \sum w_i x_i^2 y_i$$

This set of equations can be written in matrix form as

$$\mathbf{AB} = \mathbf{C} \tag{34}$$

where

$$\mathbf{A} = \begin{pmatrix} \sum w_i & \sum w_i x_i & \sum w_i x_i^2 \\ \sum w_i x_i & \sum w_i x_i^2 & \sum w_i x_i^3 \\ \sum w_i x_i^2 & \sum w_i x_i^3 & \sum w_i x_i^4 \end{pmatrix}$$

$$\mathbf{B} = \begin{pmatrix} b_0 \\ b_1 \\ b_2 \end{pmatrix}$$

$$C = \begin{pmatrix} \Sigma\, w_i y_i \\ \Sigma\, w_i x_i y_i \\ \Sigma\, w_i x_i^2 y_i \end{pmatrix}$$

To solve this set of linear equations, it is necessary to take the inverse of the coefficient matrix A. Give this new matrix the name D and call its elements d_{ij}. Then, by definition,

$$B = A^{-1}C = DC \qquad (35)$$

The values in the B matrix, namely b_0, b_1, and b_2, are the least-squares best values.

The standard deviation of the fit, s, is calculated from the equation

$$s^2 = \frac{Q}{n - p - 1} \qquad (36)$$

where Q comes from Equation (29), n is the number of data points being fit, and p is the degree of the polynomial. Note that Equation (36) reduces to Equation (21) for the linear equation where $p = 1$. Now the standard deviations for each of the parameters can be computed in terms of s^2 as

$$s_{b_0}^2 = d_{11}s^2 \qquad s_{b_1}^2 = d_{22}s^2 \qquad s_{b_2}^2 = d_{33}s^2 \qquad (37)$$

where the d_{kk} values are the elements from the main diagonal of the inverse matrix D.

These equations that have been derived for the quadratic equation are applicable for higher order polynomials as well. In general, one must invert a $(p + 1) \times (p + 1)$ matrix, where p is the degree of the polynomial. The equations for the standard deviations of the fit parameters also generalize to higher order.

A word of caution is in order to finish this discussion of polynomial curve fitting. Many computer center libraries and many software packages available to users contain routines that will take a set of data and fit polynomial equations to it in an iterative way. That is, the degree of the polynomial is increased by 1 after each fit, and the fitting is done again. Comparative statistics for the degree of goodness of the fit as a function of the polynomial degree are output. The user must carefully avoid using such routines incorrectly because it is common that a higher order polynomial will provide a better quality fit than a lower order one, but without additional true meaning. Just because a higher order polynomial can fluctuate more, it can sometimes fit noisy data better than a lower order equation. A rule of thumb that is widely practiced is to keep the degree of the polynomial being fit to less than one-third or even one-fifth of the number of data points being fit. Thus, a data set of 20 points should probably not be fit with more than about a cubic equation to be on the safe side.

3.3 LINEARIZING TRANSFORMATIONS

The equations to be fit to data are not always polynomials, let alone straight lines. Many other functional forms arise in scientific investigations, and they must be dealt with. One approach for dealing with some functional forms is to use a transformation to make the curve-fitting problem into a linear one.

An example function that can be fit using this strategy is the exponential function

$$f(x_i) = y_i = \alpha e^{\beta x_i} \tag{38}$$

If $\beta > 0$, this is a rising exponential curve (like a plot of the earth's population versus time), and if $\beta < 0$, the curve is a decaying exponential (like a plot of radioactive count rate versus time for a decaying radioactive species).

Taking natural logarithms of both sides of Equation (38) yields

$$y_i' = \ln y_i = \ln \alpha + \beta x_i \tag{39}$$

which is a linear equation after the substitutions $y_i = \ln y_i$, $a_0 = \ln \alpha$, and $a_1 = \beta$. Now

$$y_i' = a_0 + a_1 x_i \tag{40}$$

In order to maintain the proper relationship between the weights and the points, we must also transform the weights. The general equation for propagation of error is

$$\sigma_y^2 = \sum_i \sigma_{x_i}^2 \left(\frac{\partial y}{\partial x_i} \right)^2 \tag{41}$$

for any function $y = f(x_i)$. Now, the weights can normally be considered inversely related to the standard deviations of the points

$$w_i = \frac{1}{\sigma_{y_i}^2} \tag{42}$$

Then the weights are transformed using the general error propagation equation above to get

$$w_i' = w_i \left(\frac{\partial y_i'}{\partial y_i} \right)^{-2} \tag{43}$$

and

$$\frac{\partial y_i'}{\partial y_i} = \frac{1}{y_i} \tag{44}$$

so

$$w_i' = w_i y_i^2 \tag{45}$$

The transformations from the original data points (x_i, y_i) with associated weights w_i to the new linear problem are

$$y_i' = \ln y_i \quad \text{and} \quad w_i' = w_i y_i^2 \tag{46}$$

Using the transformed points and weights, a linear fitting routine will give values for the linear parameters a_0, a_1, s_{a_0}, and s_{a_1}. Back transformations are then needed to get the exponential parameters:

$$\alpha = \exp(a_0) \qquad \beta = a_1$$

$$s_\alpha^2 = \alpha s_{a_0}^2 \qquad s_\beta^2 = s_{a_1}^2$$

Thus, the exponential function of Equation (38) has been fit to a set of data using a linear least-squares routine.

EXAMPLE OF EXPONENTIAL FUNCTION FIT

An example of an exponential relationship is that of the dependence of a rate constant on the activation energy of the reaction. Measurement of the rates of a reaction as a function of temperatures allows the calculation of the $\Delta H^\dagger$ of a reaction as in this example.

The following data were reported for the temperature dependence of the rate constant k_1 for the slow step in the oxidation of Fe^{2+} by hydrogen peroxide. The equation

$$k_1 = A \exp\left(\frac{-\Delta H^\dagger}{RT}\right) \tag{47}$$

can be fit to these data to develop values for A (the preexponential factor) and $\Delta H^\dagger$ (activation energy) along with their errors. Activation energy $\Delta H^\dagger$ should be in units of calories per mole and T is in degrees Kelvin.

Temperature, °C	k_1, l mole^{-1} sec^{-1}
15.1	40.5 ± 2.0
18.6	48.9 ± 2.5
19.75	52.8 ± 2.5
20.0	51.0 ± 2.5
20.6	58.8 ± 3.0
23.8	64.4 ± 3.0
25.1	63.4 ± 3.0
29.1	81.2 ± 4.0
33.7	102.8 ± 5.0
35.5	102.5 ± 5.0
40.0	125.2 ± 6.0

The program to perform this (program EXPFIT) fitting consists of three parts: the main program, which inputs the data and calls the subroutines where the work is done; subroutine LLS, which does the least-squares curve fitting needed; and subroutine TRANS, which does the transformation of the data and the retransformations of the parameters.

```
      PROGRAM EXPFIT
C....
C.... EXPONENTIAL FUNCTION FIT BY TRANSFORMATION
C....
      DIMENSION W(20),X(20),Y(20),YP(20),WP(20),WS(20)
      COMMON /IOUNIT/ NINP,NOUT
      NINP=1
      NOUT=1
      WRITE (NOUT,1)
    1 FORMAT (' THIS IS PROGRAM EXPFIT',/)
      WRITE (NOUT,2)
    2 FORMAT (' ENTER NUMBER OF POINTS TO BE FIT')
      READ (NINP,*) N
      DO 5 I=1,N
      WRITE (NOUT,4) I
    4 FORMAT (' ENTER POINT',I3,' AND WEIGHT AS (X,Y,W)')
    5 READ (NINP,*) X(I),Y(I),WS(I)
      DO 7 I=1,N
    7 X(I)=1.0/(X(I)+273.16)
      DO 100 II=1,2
      IF (II.EQ.2) GO TO 11
      DO 10 I=1,N
   10 W(I)=1.0
      GO TO 31
   11 DO 20 I=1,N
   20 W(I)=1.0/WS(I)**2
   31 WRITE (NOUT,32)
   32 FORMAT (' ',/,10X,'X',12X,'Y',9X,'W',/)
      WRITE (NOUT,52) (X(I),Y(I),W(I),I=1,N)
   52 FORMAT (' ',F14.6,F11.2,F10.3)
      CALL TRANS (X,Y,YP,W,WP,N,AZ,SAZ,A1,SA1,AL,SAL,BE,SBE,-1)
      CALL LLS (W,X,YP,N,A1,AZ,SA1,SAZ,STD)
      CALL TRANS (X,Y,YP,W,WP,N,AZ,SAZ,A1,SA1,AL,SAL,BE,SBE,1)
      CALL LLS (WP,X,YP,N,A1,AZ,SA1,SAZ,STD)
      CALL TRANS (X,Y,YP,W,WP,N,AZ,SAZ,A1,SA1,AL,SAL,BE,SBE,1)
      BE=BE*1.9869
      SBE=SBE*1.9869
      WRITE (NOUT,55) AL,SAL,BE,SBE
   55 FORMAT (' ',//,' PREEXP. FACTOR: ',E12.3,' +/-',E12.3,//,
     X  1X,'DELTA H: ',F12.2,' +/-',F12.2)
  100 CONTINUE
      STOP
      END
C------------------- -----------------------------------------
      SUBROUTINE LLS (W,X,Y,N,A1,AZ,SA1,SAZ,STD)
C....
C.... LINEAR LEAST SQUARES FITTING ROUTINE WITH WEIGHTS
C....
      DIMENSION W(N),X(N),Y(N)
      WW=0.0
      WX=0.0
      WY=0.0
      WXY=0.0
      WXX=0.0
      WYY=0.0
      DO 10 I=1,N
      AW=W(I)
      AX=X(I)
      AY=Y(I)
```

```
      WW=WW+AW
      WX=WX+AW*AX
      WY=WY+AW*AY
      WXY=WXY+AW*AX*AY
      WXX=WXX+AW*AX*AX
   10 WYY=WYY+AW*AY*AY
      DENOM=WW*WXX-WX*WX
      A1=(WW*WXY-WX*WY)/DENOM
      AZ=(WXX*WY-WX*WXY)/DENOM
      VSUM=0.0
      DO 20 I=1,N
   20 VSUM = VSUM + W(I) * (Y(I)-AZ-A1*X(I))**2
      SS=VSUM/(N-2)
      STD=SQRT(SS)
      SA1=SQRT(SS*WW/DENOM)
      SAZ=SQRT(SS*WXX/DENOM)
      RETURN
      END
C----------- --------------------------------------------------
      SUBROUTINE TRANS (X,Y,YP,W,WP,N,AZ,SAZ,A1,SA1,AL,SAL,BE,SBE,KDUM)
C....
C.... TRANSFORMATION ROUTINE : EXPONENTIAL TO LINEAR
C....
      DIMENSION X(N),Y(N),YP(N),W(N),WP(N)
      COMMON /IOUNIT/ NINP,NOUT
      IF (KDUM) 1,1,51
    1 DO 10 I=1,N
      YP(I)=ALOG(Y(I))
   10 WP(I)=W(I)*Y(I)**2
      RETURN
   51 AL=EXP(AZ)
      SAL=SAZ*AL
      BE=A1
      SBE=SA1
      WRITE (NOUT,55) AZ,SAZ,A1,SA1,AL,SAL,BE,SBE
   55 FORMAT (' ',/,' LINEAR PARAMETERS',/,' AO',E11.4,' +/-',E11.4,/,
     X   ' A1',E11.4,' +/-',E11.4)
      WRITE (NOUT,56) AL,SAL,BE,SBE
   56 FORMAT (' EXPONENTIAL PARAMETERS',/,' AL',E11.4,' +/-',E11.4,/,
     X   ' BE',E11.4,' +/-',E11.4)
      VSUM=0.0
      DO 70 I=1,N
   70 VSUM = VSUM + W(I) * (Y(I)-AL*EXP(BE*X(I)))**2
      STD=SQRT(VSUM/FLOAT(N-2))
      WRITE (NOUT,75) STD
   75 FORMAT (' OVERALL STANDARD DEVIATION',F10.3)
      RETURN
      END


      THIS IS PROGRAM EXPFIT

      ENTER NUMBER OF POINTS TO BE FIT
      11
      ENTER POINT  1 AND WEIGHT AS (X,Y,W)
      15.1  40.5  2.0
      ENTER POINT  2 AND WEIGHT AS (X,Y,W)
      18.6  48.9  2.5
      ENTER POINT  3 AND WEIGHT AS (X,Y,W)
      19.75 52.8  2.5
      ENTER POINT  4 AND WEIGHT AS (X,Y,W)
      20.0  51.0  2.5
      ENTER POINT  5 AND WEIGHT AS (X,Y,W)
      20.6  58.8  3.0
      ENTER POINT  6 AND WEIGHT AS (X,Y,W)
      23.8  64.4  3.0
      ENTER POINT  7 AND WEIGHT AS (X,Y,W)
      25.1  63.4  3.0
```

```
 ENTER POINT  8 AND WEIGHT AS (X, Y, W)
29. 1  81. 2  4. 0
 ENTER POINT  9 AND WEIGHT AS (X, Y, W)
33. 7 102. 8  5. 0
 ENTER POINT 10 AND WEIGHT AS (X, Y, W)
35. 5 102. 5  5. 0
 ENTER POINT 11 AND WEIGHT AS (X, Y, W)
40. 0 125. 2  6. 0
```

X	Y	W
0. 003469	40. 50	1. 000
0. 003427	48. 90	1. 000
0. 003414	52. 80	1. 000
0. 003411	51. 00	1. 000
0. 003404	58. 80	1. 000
0. 003367	64. 40	1. 000
0. 003353	63 40	1. 000
0. 003308	81. 20	1. 000
0. 003259	102. 80	1. 000
0. 003240	102. 50	1. 000
0. 003193	125. 20	1. 000

```
LINEAR PARAMETERS
AO 0. 1777E+02 +/- 0. 5047E+00
A1-0. 4047E+04 +/- 0. 1506E+03

LINEAR PARAMETERS
AO 0. 5220E+08 +/- 0. 2635E+08
A1-0. 4047E+04 +/- 0. 1506E+03

EXPONENTIAL PARAMETERS
AL 0. 5220E+08 +/- 0. 2635E+08
BE-0. 4047E+04 +/- 0. 1506E+03

OVERALL STANDARD DEVIATION    3. 003

LINEAR PARAMETERS
AO 0. 1751E+02 +/- 0. 4629E+00
A1-0. 3966E+04 +/- 0. 1407E+03

LINEAR PARAMETERS
AO 0. 4009E+08 +/- 0. 1856E+08
A1-0. 3966E+04 +/- 0. 1407E+03

EXPONENTIAL PARAMETERS
AL 0. 4009E+08 +/- 0. 1856E+08
BE-0. 3966E+04 +/- 0. 1407E+03

OVERALL STANDARD DEVIATION    2. 970


PREEXP. FACTOR:   0. 401E+08 +/-   0. 186E+08

DELTA H:    -7880. 32 +/-     279. 47
```

X	Y	W
0. 003469	40. 50	0. 250
0. 003427	48. 90	0. 160
0. 003414	52. 80	0. 160
0. 003411	51. 00	0. 160
0. 003404	58. 80	0. 111
0. 003367	64. 40	0. 111
0. 003353	63. 40	0. 111
0. 003308	81. 20	0. 063
0. 003259	102. 80	0. 040
0. 003240	102. 50	0. 040
0. 003193	125. 20	0. 028

```
LINEAR PARAMETERS
AO  0. 1804E+02 +/- 0. 6282E+00
A1-0. 4129E+04 +/- 0. 1850E+03

LINEAR PARAMETERS
AO  0. 6860E+08 +/- 0. 4309E+08
A1-0. 4129E+04 +/- 0. 1850E+03

EXPONENTIAL PARAMETERS
AL  0. 6860E+08 +/- 0. 4309E+08
BE-0. 4129E+04 +/- 0. 1850E+03

OVERALL STANDARD DEVIATION    0. 856

LINEAR PARAMETERS
AO  0. 1780E+02 +/- 0. 4996E+00
A1-0. 4055E+04 +/- 0. 1492E+03

LINEAR PARAMETERS
AO  0. 5356E+08 +/- 0. 2676E+08
A1-0. 4055E+04 +/- 0. 1492E+03

EXPONENTIAL PARAMETERS
AL  0. 5356E+08 +/- 0. 2676E+08
BE-0. 4055E+04 +/- 0. 1492E+03

OVERALL STANDARD DEVIATION    0. 844

PREEXP. FACTOR:   0. 536E+08 +/-    0. 268E+08

DELTA H:    -8056. 58 +/-      296. 38
```

In addition to the exponential, other functional forms can also be transformed. An example is the hyperbolic function

$$y = \frac{x}{\alpha x - \beta} \tag{48}$$

If the transformations $y' = 1/y$ and $x' = 1/x$ are used, then the linear equation

$$y' = \alpha - \beta x \tag{49}$$

results.

The hyperbolic Equation (48) is of importance because it is the functional form that explains the relationship between the velocity of an enzymatic reaction, v, and substrate concentration s.

For enzyme reactions, the rate of reaction depends on the affinity of the substrate and enzyme. If the mechanism follows the idealized model, that is, if [substrate] $\gg$ [enzyme], and if

$$E + S \rightleftharpoons ES \rightleftharpoons E + P \tag{50}$$

then a plot of substrate concentration versus the reaction velocity produces a

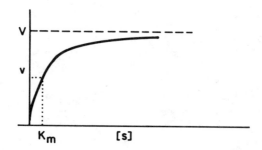

Figure 3. A plot of the Michaelis–Menten relationship between reaction velocity v and substrate concentration s.

plot of a rectangular hyperbola through the origin. The relationship between the velocity of a reaction, v, and the substrate concentration s can be expressed as the Michaelis–Menten equation:

$$v = \frac{s}{s + K_m} V \qquad (51)$$

in which K_m and V are the Michaelis–Menten constants. Here V is the velocity of the reaction when the enzyme is completely saturated with substrate and the reaction is proceeding at the maximum rate possible, and K_m is the substrate concentration at one-half the maximum velocity. A plot of the Michaelis–Menten equation is given in Figure 3. The Michaelis–Menten equation can be rearranged into the general hyperbolic Equation (48) using the substitutions $\alpha = 1/V$ and $\beta = -K_m/V$.

The Michaelis–Menten equation can be rearranged in several different ways to produce several different linear forms, as shown in the following paragraphs.

1. *Lineweaver–Burk.* The hyperbolic form of Equation (51) can be linearized by the following sequence of rearrangements: Take Equation (51), multiply both sides by $s + K_m$, divide by V, divide by v, and the equation becomes

$$\frac{1}{v} = \frac{1}{V} + \frac{K_m}{sV} \qquad (52)$$

A plot of this form is given in Figure 4, where $1/v$ is plotted against $1/s$. This linearization has been used as a basis for a graphical method for the determination of the Michaelis–Menten parameters. The values for V and K_m can be read directly off the graph as shown in the figure.

Garfinkel (1980) states, "Well over 90% of all published work in enzyme kinetics involves the use of this plot." However, Garfinkel goes on to say,

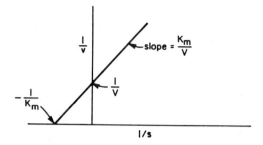

Figure 4. A double-reciprocal Lineweaver–Burk plot of the Michaelis–Menten relationship.

"By now there seems to be a consensus that, of the various linearizations of the Michaelis–Menten equation that have been proposed, this is the worst." This is because the data points with small values that are inaccurate get too much weight when the reciprocals are taken.

   2. *Michaelis–Menten.* Take Equation (51), multiply both sides by $s + K_m$, divide both sides by vV, to get

$$\frac{s}{v} = \frac{s}{V} + \frac{K_m}{V} \tag{53}$$

A plot of this form is given in Figure 5, where s/v is plotted versus s. Again, the slope and intercept of the graphical plot can be used to get values for V and K_m.

   3. *Eadie–Hofstee.* Take Equation (51), multiply both sides by $s + K_m$, divide by s, to get

$$v = V - \frac{vK_m}{s} \tag{54}$$

A plot of this form is shown in Figure 6, where v is plotted versus v/s. Again, the values for V and K_m can be obtained from the slope and intercept of the plot.

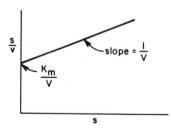

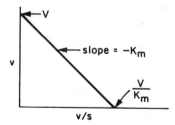

Figure 5. A plot of the Michaelis–Menten linearization of the Michaelis–Menten relationship.

Figure 6. A plot of the Eadie–Hofstee linearization of the Michaelis–Menten relationship.

EXAMPLE OF MICHAELIS–MENTEN FIT

The following is an example set of enzyme kinetic data taken from the literature (Atkinson et al. 1961).

s^a	v^b
0.138	0.148
0.220	0.171
0.291	0.234
0.560	0.324
0.766	0.390
1.460	0.493

[a] Concentration of nicotinamide mononucleotide, mM.
[b] Micromoles of nicotinamide-adenine dinucleotide formed.

Program MICMEN inputs the set of data, and then it evaluates the slope, intercept, V, and K_m using the three linearizations of the Michaelis–Menten equation. Quite different sets of values for K_m and V are obtained from these three (nominally) identical treatments.

```
      PROGRAM MICMEN
C....
C....  MICHAELIS-MENTEN CONSTANTS FROM KINETIC DATA
C....
      DIMENSION S(20),V(20),SP(20),VP(20),W(20)
      COMMON /IOUNIT/ NINP,NOUT
      NINP=1
      NOUT=1
C....
      WRITE (NOUT,1)
    1 FORMAT (' THIS IS PROGRAM MICMEN',/)
      WRITE (NOUT,2)
    2 FORMAT (' ENTER NUMBER OF POINTS TO BE FIT')
      READ (NINP,*) NN
      DO 10 I=1,NN
      WRITE (NOUT,3) I
    3 FORMAT (' ENTER POINT',I2,'  (S,V)')
   10 READ (NINP,*) S(I),V(I)
      WRITE (NOUT,11) (S(I),V(I),I=1,NN)
   11 FORMAT (' ',//,8X,'S',10X,'V',/,(' ',2F11.3))
C....  LINEWEAVER-BURK METHOD
      DO 20 J=1,NN
      W(J)=1.0
      SP(J)=1.0/S(J)
   20 VP(J)=1.0/V(J)
      CALL LLS (W,SP,VP,NN,SLOPE,AINT,SA1,SAZ,STD)
      VV=1.0/AINT
      AKM=SLOPE*VV
      WRITE (NOUT,49) VV,AKM
C....  MICHAELIS-MENTEN METHOD
      DO 30 K=1,NN
      W(K)=1.0
      SP(K)=S(K)
   30 VP(K)=S(K)/V(K)
      CALL LLS (W,SP,VP,NN,SLOPE,AINT,SA1,SAZ,STD)
      VV=1.0/SLOPE
      AKM=AINT*VV
      WRITE (NOUT,49) VV,AKM
```

```
C.... EADIE-HOFSTEE METHOD
      DO 40 L=1,NN
      W(L)=1.0
      SP(L)=V(L)/S(L)
   40 VP(L)=V(L)
      CALL LLS (W,SP,VP,NN,SLOPE,AINT,SA1,SAZ,STD)
      SLOPE=-SLOPE
      WRITE (NOUT,49) AINT,SLOPE
C....
   49 FORMAT (' ',//,5X,'MICHAELIS-MENTEN CONSTANTS',//,5X,'V =',F13.4
     1   ,/,5X,'K(M) =',F10.4)
      STOP
      END




      SUBROUTINE LLS (W,X,Y,N,A1,AZ,SA1,SAZ,STD)
C....
C.... LINEAR LEAST SQUARES FITTING ROUTINE WITH WEIGHTS
C....
      DIMENSION W(N),X(N),Y(N)
      WW=0.0
      WX=0.0
      WY=0.0
      WXY=0.0
      WXX=0.0
      WYY=0.0
      DO 10 I=1,N
      AW=W(I)
      AX=X(I)
      AY=Y(I)
      WW=WW+AW
      WX=WX+AW*AX
      WY=WY+AW*AY
      WXY=WXY+AW*AX*AY
      WXX=WXX+AW*AX*AX
   10 WYY=WYY+AW*AY*AY
      DENOM=WW*WXX-WX*WX
      A1=(WW*WXY-WX*WY)/DENOM
      AZ=(WXX*WY-WX*WXY)/DENOM
      VSUM=0.0
      DO 20 I=1,N
   20 VSUM = VSUM + W(I) * (Y(I)-AZ-A1*X(I))**2
      SS=VSUM/(N-2)
      STD=SQRT(SS)
      SA1=SQRT(SS*WW/DENOM)
      SAZ=SQRT(SS*WXX/DENOM)
      RETURN
      END




      THIS IS PROGRAM MICMEN

      ENTER NUMBER OF POINTS TO BE FIT
      6
      ENTER POINT 1   (S,V)
      0.138 0.148
      ENTER POINT 2   (S,V)
      0.220 0.171
      ENTER POINT 3   (S,V)
      0.291 0.234
      ENTER POINT 4   (S,V)
      0.560 0.324
      ENTER POINT 5   (S,V)
      0.766 0.390
      ENTER POINT 6   (S,V)
      1.460 0.493
```

```
    S           V
  0. 138      0. 148
  0. 220      0. 171
  0. 291      0. 234
  0. 560      0. 324
  0. 766      0. 390
  1. 460      0. 493
```

MICHAELIS-MENTEN CONSTANTS

```
V =         0. 5853
K(M) =      0. 4406
```

MICHAELIS-MENTEN CONSTANTS

```
V =         0. 6848
K(M) =      0. 5821
```

MICHAELIS-MENTEN CONSTANTS

```
V =         0. 6262
K(M) =      0. 4896
```

REFERENCES

M. R. Atkinson, J. F. Jackson, and R. K. Morton, "Nicotinamide Mononucleotide Adenyltransferase of Pig-Liver Nuclei," *Biochem. J.*, **80**, 318–323 (1961).

P. R. Bevington, *Data Reduction and Error Analysis for the Physical Sciences*, McGraw-Hill, New York, 1969.

S. Chaterjee and B. Price, *Regression Analysis by Example*, Wiley, New York, 1977.

A. Cornish-Bowden and R. Eisenthal, "Statistical Considerations in the Estimation of Enzyme Kinetic Parameters by the Direct Linear Plot and Other Methods," *Biochem. J.*, **139**, 721–730 (1974).

N. R. Draper and H. Smith, *Applied Regression Analysis*, 2nd ed., Wiley, New York, 1981.

H. B. Dunford, "Equilibrium Binding and Steady-State Enzyme Kinetics," *J. Chem. Ed.*, **61**, 129–132 (1984).

R. Eisenthal and A. Cornish-Bowden, "The Direct Linear Plot. A New Graphical Procedure for Estimating Enzyme Kinetic Parameters," *Biochem. J.*, **139**, 715–720 (1974).

D. Garfinkel, "Computer Modeling, Complex Biological Systems, and Their Simplifications," *Amer. J. Physiol.*, **239**, R1–R6 (1980).

G. N. Wilkinson, "Statistical Estimations in Enzyme Kinetics," *Biochem. J.*, **80**, 324–332 (1961).

4

MULTIPLE LINEAR
REGRESSION ANALYSIS

4.1 THE BASIC METHOD

In multiple linear regression analysis, we have a set of n observations, each represented by p independent variables, $x_1, x_2, \ldots, x_p$. The observation number is given the subscript i and the variable number is given the subscript j. Thus, the jth independent variable for the ith observation is denoted x_{ij} and the ith dependent variable y_i. The mathematical model that relates the y values to the x values is assumed to be linear and of the form

$$\hat{y}_i = a_0 + a_1 x_1 + a_2 x_2 + \cdots + a_p x_p \tag{1}$$

where the $a_1, a_2, \ldots, a_n$ are the regression coefficients and $\hat{y}_i$ is the predicted value of the ith observation. We call this formulation multiple linear regression because y is a linear function of x. The residual is defined as the difference between the observed and predicted values for the ith observation:

$$e_i = y_i - \hat{y}_i \tag{2}$$

It is assumed that the residuals are normally distributed, random variables. That is, the residuals contain no useful information about the relationships that might exist between the y values and the x values. Thus, they should be uncorrelated with these and with each other as well.

In multiple linear regression, the values of the coefficients are found by the same general procedure that has been discussed earlier with reference to simple linear regression. The sum of the squared residuals,

$$Q = \sum_{i=1}^{n} e_i^2 \tag{3}$$

is minimized by using calculus.

The best estimates of the coefficients, which are those values that minimize Q, are the solutions of the following set of normal equations:

$$S_{11}a_1 + S_{12}a_2 + \cdots + S_{1p}a_p = S_{y1}$$

$$S_{12}a_1 + S_{22}a_2 + \cdots + S_{2p}a_p = S_{y2}$$

$$\vdots$$

$$S_{1p}a_1 + S_{p2}a_2 + \cdots + S_{pp}a_p = S_{yp} \tag{4}$$

where the summations within the equations are defined as

$$S_{ij} = \sum_{k=1}^{n} (x_{ik} - \bar{x}_i)(x_{jk} - \bar{x}_j) \qquad i, j = 1, 2, \ldots, p \tag{5}$$

$$S_{yi} = \sum_{k=1}^{n} (y_k - \bar{y})(x_{ik} - \bar{x}_i) \qquad i = 1, 2, \ldots, p \tag{6}$$

$$\bar{x}_i = \frac{1}{n} \sum_{k=1}^{n} x_{ik} \tag{7}$$

$$\bar{y} = \frac{1}{n} \sum_{k=1}^{n} y_k \tag{8}$$

and the constant coefficient is calculated as

$$a_0 = \bar{y} - a_1\bar{x}_1 - a_2\bar{x}_2 - \cdots - a_p\bar{x}_p \tag{9}$$

The normal equations can be solved by the usual methods for solving systems of linear equations, and this operation is not relevant here.

The overall standard deviation of the fit is given by

$$s^2 = \frac{Q}{n - p - 1} \tag{10}$$

where Q is the sum-squared error (SSE) defined above. The coefficients, the a_j's, have standard deviations given by

$$s_{a_j} = (s^2 c_{jj})^{1/2} \tag{11}$$

where c_{jj} is the jth element on the main diagonal of the inverse of the corrected sum of squares and products matrix.

In matrix notation, the matrix containing the independent variables X and the matrix containing the dependent variables Y are

$$\mathbf{X} = \begin{pmatrix} x_{01} & x_{11} & \cdots & x_{p1} \\ x_{01} & x_{12} & \cdots & x_{p2} \\ \vdots & \vdots & \ddots & \vdots \\ x_{0n} & x_{1n} & \cdots & x_{pn} \end{pmatrix} \qquad \mathbf{Y} = \begin{pmatrix} y_1 \\ y_2 \\ \vdots \\ y_n \end{pmatrix} \tag{12}$$

Then, if the coefficients to be found are

$$\mathbf{a} = \begin{pmatrix} a_1 \\ a_2 \\ \vdots \\ a_p \end{pmatrix} \tag{13}$$

the least-squares estimate of the coefficients can be expressed as

$$\mathbf{a} = (\mathbf{X'X})^{-1}\mathbf{X'Y} \tag{14}$$

Then if the matrix $(\mathbf{X'X})^{-1}$ is given the name $\mathbf{C}$, its main diagonal elements, c_{jj}, appear in the equation above.

The multiple correlation coefficient R is a measure of the adequacy of the fit to the data. It is defined as

$$R^2 = 1 - \frac{\Sigma(y_i - \hat{y}_i)^2}{\Sigma(y_i - \bar{y})^2} \tag{15}$$

It has values between 0 and 1. Values of R^2 near 1 occur when y_i and $\bar{y}_i$ are nearly identical, which means that the model can reproduce the y_i values well. Values of R^2 near zero mean that the model is not a good one.

Reduced Model

The model of Equation (1) is the full model because it uses all of the available independent variables. However, if only a subset of the independent variables are used to construct a model, the result will be called a reduced model. Any reduced model can be compared to the full model in a formal way as follows.

The symbol $\hat{y}_i$ represents the values predicted for y_i by the full model. Let $\hat{y}_i^*$ be the value of y_i predicted by the reduced model. The sum-squared error of the full model is

$$\text{SSE}_{\text{full}} = \Sigma(y_i - \hat{y}_i)^2 \tag{16}$$

and the SSE for the reduced model is

$$\text{SSE}_{\text{red}} = \Sigma(y_i - \hat{y}_i^*)^2 \tag{17}$$

While the full model contains p parameters, the reduced model has q parameters. The appropriate test to determine whether the reduced model represents the data just as well as the full model is to compare the SSEs. We must correct for the fact that different numbers of parents are involved, so we compare

$$\frac{SSE_{red} - SSE_{full}}{p + 1 - q} \tag{18}$$

and

$$\frac{SSE_{full}}{n - p - 1} \tag{19}$$

The ratio of these quantities is given by

$$F = \frac{(SSE_{red} - SSE_{full})(n - p - 1)}{(p + 1 - q)(SSE_{full})} \tag{20}$$

This ratio has the F-distribution with $p + 1 - q$ and $n - p - 1$ degrees of freedom.

The F-value can be computed for the full model and any reduced model, and the value of F obtained can be compared to the F-values in tables to determine whether the reduced model represents the data as well as the full model. This is the fundamental method used by multiple linear regression analysis routines to compare two models.

While the value of R^2 is a good guide to the quality of a model, other questions about a model should also be asked. Several important questions to raise are as follows: Are all the regression coefficients significantly different from zero? Are some of the regression coefficients zero? Are some of the regression coefficients equal to each other? All these questions can be answered by using the approach culminating in the computation of the F-value of two models and comparing them. In addition, the individual coefficients in the equation have associated F-values, and these can be used to assess the significance in the context of the other variables in the model.

Variable Selection: Stepwise Procedures

Given a data set containing n observations and many independent variables per observation, the question often arises as to whether some subset of the variables will support a good model. That is, which of the available variables should actually be used in the model? This is variable selection.

Two models, one with p variables and one with q variables, can be compared according to many statistical criteria. They all rest ultimately on computing the SSEs for the two models and comparing them. A list of criteria that have been suggested is available (Hocking 1976). One criterion is the residual mean square. For an equation with p terms, it is

$$RMS = \frac{SSE}{n - p} \tag{21}$$

When comparing two equations, the one with the smaller value for RMS would be preferred. The RMS is related to the multiple correlation coefficient as

$$R^2 = 1 - \frac{(n-p)\text{RMS}}{\Sigma\,(y_i - \bar{y})^2}$$

$$= 1 - \frac{\text{SSE}}{\text{SST}} \tag{22}$$

One way to determine which equation is best for a set of variables is to fit all possible subset equations to the data. If there are n variables, there are 2^n equations to be fit. While this method is workable for small values of n, it becomes impractical as n becomes large.

An attractive alternative is to introduce or delete variables one at a time. The *forward selection* procedure starts with just one variable and then selects the best of those remaining to use second, and so on, until all the variables have been entered. The definition of *best* used could be the RMS or R^2. The *backward selection* procedure starts with the full model and deletes one variable at a time until none are left. The variable to be deleted at each stage is the one that contributes the least to the reduction of the SSE. The *stepwise* method combines these two approaches. It starts with the one best variable and then adds another. After each addition of a new variable, all the variables currently in the model are considered as candidates for deletion. Thus, a variable added to the model earlier can be deleted later in the selection process. The various stepwise multiple linear regression methods may give different final results for the same set of data. Many statistical software packages (e.g., SAS, SPSS, BMDP, Minitab) include stepwise multiple linear regression analysis routines.

4.2 IMPLEMENTATION OF MULTIPLE LINEAR REGRESSION ANALYSIS

The SWEEP Operator

Many statistical computations can be done using the SWEEP operator (Beaton 1964, Goodnight 1979, Jennrich 1979). Operationally, it is defined as follows.

Given a symmetric positive definite matrix $\mathbf{A}$:

1. Set $D = a_{kk}$.
2. Divide row k by D.
3. For all rows $i = k$, set $B = a_{ik}$ and subtract B times row k from row i. Set $a_{ik} = -B/D$.
4. Set $a_{kk} = 1/D$.

Specifically, the SWEEP operator can be used to carry out the computation needed in multiple linear regression.

Program MREG

Program MREG is an interactive FORTRAN program to perform multiple linear regression. It steps forward adding one variable at a time, printing out the results of each step.

The program is interactive, with all the input coming through the interactive terminal and all output going there. The main program is a framework for calling the subroutines that actually perform all of the computations. The routine named DATARD inputs that data, sets up the matrices needed for the remainder of the operations, and prints out the means and standard deviations for each of the variables. The heart of the computation is performed by a subroutine named SWEEP that implements the SWEEP algorithm described above. It contains no input–output operations. Subroutine FINISH prints out the results of each step and the F to enter of each variable not yet included in the model.

Program MREG has been executed with a sample set of data to show how multiple linear regression works.

```
        PROGRAM MREG
C....   PROGRAM TO PERFORM MULTIPLE LINEAR REGRESSIONS
        DOUBLE PRECISION A
        DIMENSION A(52,52),SSS(51)
        COMMON /IOUNIT/ NINP,NOUT
        NINP=1
        NOUT=1
        WRITE (NOUT,1)
      1 FORMAT (' THIS IS PROGRAM MREG',/)
        CALL DATARD(IP,N,A)
        IF(IP.LE.O.OR.N.LE.O.OR.IP.GE.N) STOP
C.....SWEEP ON CONSTANT TERM
        IP1 = IP + 1
        CALL SWEEP(A,IP1,1)
C.....SST IS A(IP1,IP1)
        SST = A(IP1,IP1)
C.....FOR EACH X, CALL SWEEP, COLLECT SSS TERM
        DO 2 I = 2,IP
           SSS(I) = 0.0
           SSBEF = A(IP1,IP1)
           CALL SWEEP(A,IP1,I)
           SSAFT = A(IP1,IP1)
           SSS(I) = SSBEF-SSAFT
           CALL FINISH(A,I,IP1,SSS,SST,N)
      2 CONTINUE
        STOP
        END
C-----------------------------------------------------------------
        SUBROUTINE DATARD(IP,N,A)
C.....READS A LINE OF DATA (X1...XP,Y) ADDS XO
C....    THEN UPDATES MEAN AND STD DEV VALUES OF EACH VARIABLE
        DOUBLE PRECISION A,U,V
        DIMENSION X(52),A(52,52),XBAR(52),XSTD(52),U(52),V(52)
        COMMON /IOUNIT/ NINP,NOUT
        WRITE (NOUT,1001)
```

```
 1001 FORMAT (' ENTER TOTAL NUMBER OF VARS PER LINE (X+Y)')
      READ (NINP,*) IP
      IF(IP.LE.0) RETURN
      WRITE (NOUT,1002)
 1002 FORMAT (' ENTER NUMBER OF OBSERVATIONS')
      READ (NINP,*) N
      WRITE (NOUT,1003)
 1003 FORMAT (' ENTER X1,X2,...,XP AND Y')
      IP1 = IP + 1
C.... INITIALIZE THE ARRAYS
      DO 1005 I = 1,52
         XBAR(I) = 0.
         XSTD(I) = 0.
         U(I) = 0.D0
         V(I) = 0.D0
         DO 1004 J = 1,52
            A(I,J) = 0.D0
 1004    CONTINUE
 1005 CONTINUE
C.....READ A CASE, UPDATE ARRAYS BY PROVISIONAL MEANS
      DO 1008 KK=1,N
      READ (NINP,*) (X(I),I = 2,IP1)
      X(1) = 1.0
C.....UPDATE XBAR, XSTD, AND A, ALL IN ONE DO LOOP
      DO 1007 J = 1,IP1
         U(J) = X(J)-XBAR(J)
         XBAR(J) = (FLOAT(KK-1) * XBAR(J) + X(J)) /FLOAT(KK)
         V(J) = X(J)-XBAR(J)
         XSTD(J) = XSTD(J) + U(J) * V(J)
         DO 1006 K = 1,J
            A(K,J) = A(K,J) + X(J) * X(K)
            A(J,K) = A(K,J)
 1006    CONTINUE
 1007 CONTINUE
 1008 CONTINUE
C.....PRINT MEANS
      WRITE (NOUT,1009)
 1009 FORMAT (' MEANS: ')
      WRITE (NOUT,1010) (XBAR(I),I=2,IP1)
 1010 FORMAT (' ',5G10.4)
C.....COMPUTE AND PRINT STANDARD DEVIATIONS
      NM1 = N-1
      DO 1011 I = 2,IP1
         XSTD(I) = SQRT(XSTD(I) / NM1)
 1011 CONTINUE
      WRITE (NOUT,1012)
 1012 FORMAT (' STANDARD DEVIATIONS: ')
      WRITE (NOUT,1013) (XSTD(I),I=2,IP1)
 1013 FORMAT (' ',5G10.4)
      RETURN
      END
C-------------------- ------------------------------------
      SUBROUTINE SWEEP(A,IP1,K)
C.....SWEEPS MATRIX A ON K-TH ROW AND COLUMN
      DOUBLE PRECISION A(52,52),B,D
      D = A(K,K)
      DO 1 J = 1,IP1
         A(K,J) = A(K,J) / D
    1 CONTINUE
      DO 3 I = 1,IP1
         IF(I.EQ.K) GO TO 3
         B = A(I,K)
         DO 2 J = 1,IP1
            A(I,J) = A(I,J)-B * A(K,J)
    2    CONTINUE
         A(I,K) = -B / D
    3 CONTINUE
      A(K,K) = 1.D0 / D
      RETURN
      END
```

```
C------- -----------------------------------------------------------
      SUBROUTINE FINISH(A, IX, IP1, SSS, SST, N)
C.... PRINTS RESULTS OF STEP
C....      IX IS NUMBER OF VARIABLES IN CURRENT MODEL
      DOUBLE PRECISION A(52, 52)
      DIMENSION SSS(52)
      COMMON /IOUNIT/ NINP, NOUT
      IP = IP1-1
      IDFE = N-IX
      WRITE (NOUT, 1) IX-1
    1 FORMAT (/, ' - - - - - STEP', I3, ' - - - - -')
      WRITE (NOUT, 2) IDFE
    2 FORMAT ('    I    A(I)', 8X, 'S. E. ', 7X, 'T(', I2, ' DF)', 4X, 'SEQ. SS')
C..... MSE
      SSE = A(IP1, IP1)
      XMSE = SSE / IDFE
C..... STDEV
      S = SQRT(XMSE)
      DO 4 I = 1, IX
         BETA = A(I, IP1)
         SA = A(I, I)
         STDERR = S * SQRT(SA)
         T = BETA / STDERR
      WRITE (NOUT, 3) I-1, BETA, STDERR, T, SSS(I)
    3 FORMAT (' ', I3, 4E12. 4)
    4 CONTINUE
      SSR = SST-SSE
      RSQ = 1. -SSE / SST
      IDF1 = IX-1
      F = SSR / IDF1
      F = F / XMSE
      WRITE (NOUT, 5) RSQ, IDF1, IDFE, F
    5 FORMAT (' R-SQ =', F6. 3, 5X, ' F(', I2, ', ', I2, ' DF) = ', E12. 4)
      WRITE (NOUT, 6) N, S
    6 FORMAT (' NOBS = ', I2, 5X, 'STD DEV = ', E12. 4)
C..... FOR VARS NOT YET IN MODEL
      IF(IX. EQ. IP) RETURN
      IXP1 = IX + 1
      WRITE (NOUT, 7) IDFE-1
    7 FORMAT (' VARS NOT IN MODEL - F TO ENTER (1, ', I2, ' DF): ')
      DO 9 I = IXP1, IP
         AII = A(I, I)
         AID = ABS(A(I, IP1))
         F = AID * AID * (IDFE-1) / (AII * SSE-AID * AID)
         WRITE (NOUT, 8) I-1, F
    8    FORMAT (' ', I5, E12. 4)
    9 CONTINUE
      RETURN
      END


   THIS IS PROGRAM MREG

   ENTER TOTAL NUMBER OF VARS PER LINE (X+Y)
   3
   ENTER NUMBER OF OBSERVATIONS
   8
   ENTER X1, X2, ..., XP AND Y
   0. 1   0. 01   0. 024
   0. 2   0. 04   0. 05
   0. 3   0. 09   0. 098
   0. 4   0. 16   0. 13
   0. 5   0. 25   0. 162
   0. 6   0. 36   0. 193
   0. 8   0. 64   0. 26
   1. 0   1. 0    0. 29
```

```
MEANS:
0. 4875      0. 3187      0. 1509
STANDARD DEVIATIONS:
0. 3044      0. 3432      0. 9462E-01

- - - - - STEP  1 - - - - -
   I    A(I)            S.E.          T( 6 DF)      SEQ. SS
   0  0. 4451E-03   0. 8557E-02   0. 5202E-01   0. 0000E+00
   1  0. 3086E+00   0. 1516E-01   0. 2036E+02   0. 6177E-01
R-SQ = 0. 986          F( 1, 6 DF) =     0. 4145E+03
NOBS =   8        STD DEV =    0. 1221E-01
VARS NOT IN MODEL - F TO ENTER (1, 5 DF):
      2  0. 1300E+02

- - - - - STEP  2 - - - - -
   I    A(I)            S.E.          T( 5 DF)      SEQ. SS
   0 -0. 2458E-01   0. 8519E-02  -0. 2885E+01   0. 0000E+00
   1  0. 4361E+00   0. 3644E-01   0. 1197E+02   0. 6177E-01
   2 -0. 1166E+00   0. 3233E-01  -0. 3606E+01   0. 6458E-03
R-SQ = 0. 996          F( 2, 5 DF) =     0. 6283E+03
NOBS =   8        STD DEV =    0. 7048E-02
```

REFERENCES

A. E. Beaton, "The Use of Special Matrix Operators in Statistical Calculus," Educational Testing Service, Research Bulletin No. RB-64-51, pp. 64–51 (1964).

N. R. Draper and H. Smith, *Applied Regression Analysis*, 2nd ed., Wiley, New York, 1981.

J. H. Goodnight, "Tutorial on the SWEEP Algorithm," *Amer. Stat.*, **33**, 149–158 (1979).

R. R. Hocking, "The Analysis and Selection of Variables in Linear Regression," *Biometrics*, **32**, 1–49 (1976).

R. I. Jennrich, "Stepwise Regression," in *Statistical Methods for Digital Computers*, K. Enslein (Ed.), Wiley, New York, 1979, pp. 58–75.

G. A. F. Seber, *Multivariate Observations*, Wiley, New York, 1984.

5

NUMERICAL INTEGRATION

Integration consists of finding a function whose derivative is given. In terms of the general definite integral

$$\int_a^b f(x)\, dx \tag{1}$$

the integral's value is given by

$$F(b) - F(a) \tag{2}$$

where $f(x)$ is the first derivative of $F(x)$, that is, $F'(x) = f(x)$.

Two types of integrals are indefinite and definite. The result of solving an indefinite integral is an analytical function plus a constant of integration. That is,

$$\int f(x)\, dx = F(x) + c \tag{3}$$

and a specific example is

$$\int (3x^2 - 2)\, dx = x^3 - 2x + c \tag{4}$$

The result of solving a definite integral is a number. For the example function above, the value of the definite integral between the limits of 0 and 2 would be calculated as

$$\int_0^2 (3x^2 - 2)\, dx = x^3 - 2x + c \Big]_0^2$$

$$= (8 - 4 + c) - (0 - 0 + c)$$

$$= 4 \tag{5}$$

After evaluation of the solution function at the limits of integration, the result is a number, here 4. This particular definite integral was evaluated by first performing an indefinite integration and then evaluating the solution function $F(x)$ at the desired limits. This two-step procedure for evaluating definite integrals should always be used whenever possible because it is exact. However, when this exact, direct procedure is not applicable, a variety of techniques are available, such as integration by parts, substitutions, and transformations. However, these techniques often fail, in which case numerical evaluation approaches must be used.

Several of the situations that can lead to the necessity of using numerical integration are as follows: One or both of the integration limits can be infinite; only discrete data points might be available $(x, f(x))$ rather than a function to be integrated; or the function can be downright difficult and intractable.

The basis for all numerical integration methods is as follows. The value of the definite integral

$$\int_a^b f(x)\, dx \tag{6}$$

is given by the area under the curve $y = f(x)$ between $x = a$ and $x = b$. This is illustrated in Figure 1. This area can be estimated by subdividing the total area under the curve between a and b into subintervals whose areas can be calculated individually, as shown in Figure 2. Each of the subintervals has three sides that are straight lines and a fourth, the top, that is curved.

If the top of each subinterval were approximated with a short straight-line segment, then the area of the subinterval could be calculated exactly because it would be a trapezoid.

Although the area of the trapezoid can be calculated exactly, it is an approximation to the true area of the subinterval. The difference between the true area and the trapezoid's area is the truncation error. It will be small only if the function being integrated, $f(x)$, is itself nearly linear across the subinterval and if the subinterval is narrow.

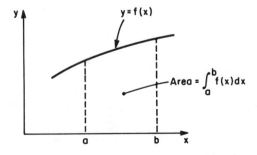

Figure 1. The definite integral as an area.

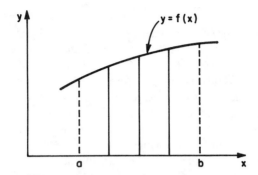

Figure 2. The definite integral as a summation of areas of subintervals.

The quality of the approximation to the integral can be improved by using a higher order polynomial to approximate the curve at the top of the subinterval. The next higher order polynomial after the straight line is the quadratic function or parabola. The specification of three points is sufficient to uniquely define a parabola, as shown in Figure 3. The three points (x_1, y_1), (x_2, y_2), (x_3, y_3) lie on the function being integrated with a spacing of h. Let the parabola passing through these three points be represented by

$$y = a_1 x^2 + a_2 x + a_3 \tag{7}$$

where the three coefficients a_1, a_2, a_3 are determined by the three specified points. The area under the parabola between $x = x_1$ and $x = x_3$ is given by

$$A = \int_{x_1}^{x_3} y \, dx = \int_{x_2-h}^{x_2+h} (a_1 x^2 + a_2 x + a_3) \, dx \tag{8}$$

Substitution of the identity

$$y_2 = a_1 x_2^2 + a_2 x_2 + a_3 \tag{9}$$

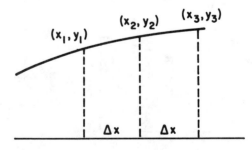

Figure 3. A parabola passing through three specified points.

into Equation (8) and algebraic rearrangement yields

$$A = 2y_2 h + \tfrac{2}{3} a_1 h_3 \tag{10}$$

We can show that

$$y_1 + y_3 = 2y_2 + 2a_1 h^2 \tag{11}$$

and use this identity to eliminate a_1 from Equation (10), yielding

$$A = \tfrac{1}{3} h(y_1 + 4y_2 + y_3) \tag{12}$$

This simple equation provides a way to compute the exact area under the parabola from $x = x_1$ to $x = x_3$ in Figure 3.

To calculate the area for the entire integral from a to b, the region from a to b is divided into n equal parts (n must be even). Then Equation (12) is applied to each pair of intervals, a summation is taken (remembering to count each subinterval only once overall), and the final equation is

$$A = \tfrac{1}{3} h[f(x_0) + 4f(x_1) + 2f(x_2) + 4f(x_3) + 2f(x_4)$$
$$+ \cdots + 4f(x_{n-1}) + f(x_n)] \tag{13}$$

This numerical integration equation is called Simpson's rule. To obtain the estimate of the integral requires evaluating the function $f(x)$ a total of $n + 1$ times, multiplying by integer coefficients, and summing the contributions.

The basis of the Simpson rule approach to numerical integration is approximating the actual function to be integrated, $f(x)$, with a parabola over each narrow subinterval. In general, this will lead to errors in the estimates because the parabola will not exactly mimic the function $f(x)$. The error cannot be computed (or we could add it to the approximation to get the exact result), but it can be estimated as follows. The error of estimating the area within a pair of subintervals as expressed by Equation (12) will govern the overall error of the procedure, so the derivation will be done on that basis. The error is given by

$$\int_{x_1}^{x_3} f(x)\, dx - \tfrac{1}{3} h(y_1 + 4y_2 + y_3) \tag{14}$$

We assume that our function $f(x)$ can be expanded as a Taylor series about the midpoint (x_2, y_2), to get

$$f(x) = y_2 + xy_2' + \frac{x^2}{2!} y_2'' + \frac{x^3}{3!} y_2''' + \cdots \tag{15}$$

Now $y_1 = f(x_1)$ and $y_3 = f(x_3)$. The first three terms of the Taylor expansion must be zero because they constitute a parabola themselves. Therefore, the first term of Equation (15) that we would expect to be positive would be the h^3 term. However, when we substitute $(x^3/3!)y_2'''$ into Equation (14) and solve, $error_3$ is zero because of the odd symmetry. The next higher-order term in the Taylor series is $(x^4/4!)y_2^{iv}$. When this is substituted into Equation (14), we get

$$error_4 \cong f^{iv}(x)h^5 \tag{16}$$

after some algebraic manipulations. The absolute magnitude of this term is not so important as its functional form. It says that the error expected from a Simpson rule numerical integration—if the Taylor series well approximates the function about the midpoint of the subinterval—goes as the fourth derivative of the function and as the fifth power of h. Thus, integrals of functions for which the fourth derivative is small are approximated with little error. But functions with sizable fourth derivatives, for example, sinusoidal or exponential functions, are not approximated as well. The equation also shows that halving the interval size should decrease the expected error by a factor of 2^{-5}, or $\frac{1}{32}$. This is quite favorable because halving the interval size only doubles the computational effort but yields $\frac{1}{32}$ the expected error.

Program NISR

Simpson's rule numerical integration can be implemented in a short, straightforward FORTRAN program, NISR, which implements Simpson's rule in an interactive routine. It integrates the equation

$$\int e^{-x}x^{a-1}\,dx \tag{17}$$

for an a value and integration limits input by the user. The function itself is in the form of a statement function using the dummy variable ZZ. To integrate a different function using this program would require changing the statement function.

Upon execution, the routine prompts the user to input a value for the parameter a, values for the lower and upper limits of integration, the number of subintervals to start the integration, and the relative accuracy desired for a stopping criterion for the iteration. The routine calculates the integral values using the Simpson rule equation, checks the relative accuracy, doubles the number of subintervals, prints a status report for the iteration, and starts over. It continues this iterative procedure until the following stopping criterion is satisfied:

$$\frac{|A_{current} - A_{last}|}{A_{current}} < \epsilon \tag{18}$$

where $A_{current}$ is the current estimate for the integral and A_{last} is the prior estimate. If the error criterion ϵ is unrealistically small, the routine will not terminate as desired but will lose precision due to round-off errors as the number of subintervals gets very large.

Equation (17) is very closely related to the gamma function, which is

$$\Gamma(a) = \int_0^\infty e^{-x} x^{a-1} \, dx \tag{19}$$

The gamma function is the extension of the factorial function to nonintegral arguments. The upper limit of integration of the gamma function is infinite, so we must approximate this upper limit while using Simpson's rule. However, this approximation will not lead to unacceptable errors because the function being integrated is dominated by the term e^{-x}, which becomes very small as x becomes large. Thus, if we use a sufficiently large value for the upper limit of integration, we will not be neglecting very much area under the curve.

The accompanying figures show two executions of NISR using upper limits of integration of 25 and 50. An ϵ value of 0.001 was used in each case. Identical values for the integral were obtained, as the output shows. The actual value of the gamma function for this value of a is $\Gamma(1.6) = 0.89352$, so this simple routine NISR obtained a value in error by 2 parts in 8900, or 0.022%. This funtion, being basically an exponential function, is not a particularly favorable case for Simpson's rule since the fourth derivative is large. Nonetheless, the Simpson rule integration generated a relatively accurate value.

```
       PROGRAM NISR
C....
C....  NUMERICAL INTEGRATION WITH SIMPSONS RULE
C....
C....  THE FUNCTION BEING INTEGRATED
C....
       FUNC(ZZ)=ZZ**AA *EXP(-ZZ)
C....
       DATA NOUT/1/,NSTART/8/,ACC/1.OE-03/
C....
C....  INPUT PARAMETERS AND PRINT OUT CONDITIONS
C....
       WRITE (NOUT,101)
  101 FORMAT (' NUMERICAL INTEGRATION WITH SIMPSONS RULE',/)
       WRITE (NOUT,102)
  102 FORMAT (' INPUT A VALUE FOR A')
       READ (NOUT,*) A
       WRITE (NOUT,103) A
  103 FORMAT (' A = ',F5.2)
       WRITE (NOUT,104)
  104 FORMAT (' INPUT LEFT AND RIGHT INTERVAL BOUNDARIES')
       READ (NOUT,*) X1,X2
       WRITE (NOUT,105) X1,X2
  105 FORMAT (' LEFT BOUNDARY =',F5.2,5X,'RIGHT BOUNDARY =',F6.2)
       WRITE (NOUT,106) NSTART
  106 FORMAT (' NUMBER OF INTERVALS AT START =',I4)
       WRITE (NOUT,107) ACC
```

```
  107 FORMAT (' RELATIVE ACCURACY REQUIRED =',F8.4)
      WRITE (NOUT,109)
  109 FORMAT (//,' NO. OF INTERVALS',5X,'INTEGRAL VALUE',/)
C....
C.... PERFORM CALCULATIONS
C....
      N=NSTART
      FA=0.0
      AA=A-1.0
      IF (X1) 4,3,4
    3 CON=FUNC(X2)
      GO TO 5
    4 CON=FUNC(X1)+FUNC(X2)
    5 CONTINUE
    6 DX=(X2-X1)/FLOAT(N)
      SUM1=0.0
      SUM2=0.0
      L=N/2-1
      X=X1
      DO 10 I=1,L
      X=X+DX
      SUM1=SUM1+FUNC(X)
      X=X+DX
   10 SUM2=SUM2+FUNC(X)
      F=(CON+4.0*SUM1+2.0*SUM2)*DX/3.0
      WRITE (NOUT,111) N,F
  111 FORMAT (' ',5X,I5,12X,F8.4)
      IF (ABS(F-FA)-ACC*ABS(FA)) 13,13,12
   12 N=2*N
      FA=F
      GO TO 6
   13 CONTINUE
      STOP
      END


      NUMERICAL INTEGRATION WITH SIMPSONS RULE

      INPUT A VALUE FOR A
      1.6
      A =   1.60
      INPUT LEFT AND RIGHT INTERVAL BOUNDARIES
      0.0 25.0
      LEFT BOUNDARY = 0.00     RIGHT BOUNDARY = 25.00
      NUMBER OF INTERVALS AT START =    8
      RELATIVE ACCURACY REQUIRED =   0.0010


         NO. OF INTERVALS        INTEGRAL VALUE

                8               0.3762
               16               0.7194
               32               0.8455
               64               0.8795
              128               0.8892
              256               0.8921
              512               0.8930
             1024               0.8933


      NUMERICAL INTEGRATION WITH SIMPSONS RULE

      INPUT A VALUE FOR A
      1.6
      A =   1.60
      INPUT LEFT AND RIGHT INTERVAL BOUNDARIES
      0.0 50.0
      LEFT BOUNDARY = 0.00     RIGHT BOUNDARY = 50.00
      NUMBER OF INTERVALS AT START =    8
      RELATIVE ACCURACY REQUIRED =   0.0010
```

NO. OF INTERVALS	INTEGRAL VALUE
8	0.0484
16	0.3762
32	0.7194
64	0.8455
128	0.8795
256	0.8892
512	0.8921
1024	0.8930
2048	0.8933

Gaussian Quadrature

To this point in our discussion of numerical integration, we have used polynomial approximations across evenly spaced intervals. If we are prepared to use unequally spaced intervals, then alternative approaches are available. In Gaussian quadrature, the subinterval boundaries are chosen so as to optimize the estimation of the area of the subintervals.

Gaussian quadrature utilizes a summation equation to approximate an integral:

$$\int_a^b w(x)f(x)\,dx \cong \sum_{k=1}^n A_k f(x_k) \tag{20}$$

The function being integrated is the product of $w(x)$, the weighting function, and $f(x)$. The coefficients A_k are called the weighting factors, and the x_k values are the abscissas at which the function $f(x)$ is to be evaluated. The symbol n is the number of terms in the summation, that is, the number of subinterval boundaries.

There are a number of different classes of Gaussian quadrature equations, depending on the lower limit of integration a, the upper limit of integration, b, and the functional form of the weighting function $w(x)$. The simplest case, and therefore the most widely applicable, is known as Gauss–Legendre quadrature. In this case, $a = -1$, $b = 1$, and $w(x) = 1$, so Equation (20) simplifies to

$$\int_{-1}^1 f(x)\,dx \cong \sum_{k=1}^n A_k f(x_k) \tag{21}$$

This quadrature equation is applicable to all functions $f(x)$, but the general limits of integration, a to b, must be transformed to the range -1 to $+1$. This requires a linear change of variable from x to t as follows:

$$t = \frac{2x - (a + b)}{b - a} \tag{22}$$

$$x = \tfrac{1}{2}(b - a)t + \tfrac{1}{2}(b + a) \tag{23}$$

Integrating t over the interval -1 to $+1$ is algebraically equal to integrating x over the interval a to b:

$$\int_a^b f(x)\,dx = \int_{-1}^1 g(t)\,dt \tag{24}$$

where

$$g(t) = f[\tfrac{1}{2}(b-a)t + \tfrac{1}{2}(b+a)] \tag{25}$$

The final equation is

$$\int_a^b f(x)\,dx \cong \frac{b-a}{2} \sum_{k=1}^n A_k f(x_k) \tag{26}$$

where the x_k values have been transformed using Equation (23).

To actually perform a Gauss–Legendre integration of a given function requires going to a mathematical handbook (Davis and Polonsky 1964) to find the values for the A_k and x_k, $k = 1, 2, 3, \ldots, n$, for the value of n chosen. Then these values are plugged into the equations above. Of course, the weighting factors A_k and the abscissa values x_k are contained in subroutines within the libraries of most large computer systems.

The following example carries through a specific case of using Gauss–Legendre quadrature. We will approximate the integral

$$\int_0^{\pi/2} x^2 \cos x\,dx \tag{27}$$

using a four-point summation, that is, $n = 4$. From tables found in Davis and Polonsky (1964), the four t_k values are ± 0.3399810436 (t_2 and t_3) and ± 0.8611363116 (t_1 and t_4) and the four A_k values are ± 0.6521451549 (for t_2 and t_3) and ± 0.3478548451 (for t_1 and t_4). The next step is to transform the abscissa values using Equation (23):

$$x_1 = \frac{(\pi/2)-0}{2} + \frac{(\pi/2)-0}{2}\,t_1 = 0.109064 \tag{28}$$

and

$$x_2 = 0.518378 \qquad x_3 = 1.05242 \qquad x_4 = 1.46173$$

Then we must compute $f(x_k) = x^2 \cos(x_k)$ for the four x_k values: $f(x_1) = 0.109064^2 \; \cos(0.109064) = 0.01824$, $f(x_2) = 0.233413$, $f(x_3) = 0.548777$, and $f(x_4) = 0.232572$. Finally,

$$I = \frac{(\pi/2)-0}{2} \sum_{k=1}^4 A_k f(x_k) = 0.467402 \tag{29}$$

The exact value for this integral is 0.467401. Thus, our approximation is off by 1 part in 467,400 or 2 parts per 1×10^6 and with only four evaluations of the function. Gauss–Legendre integration can be quite accurate.

As mentioned above, there are a number of different cases of Gaussian quadrature. Gauss–Legendre is one, but others are useful for integrating certain function types. For functions containing a term e^{-x} and integration limits of zero and infinity, the Gauss–Laguerre quadrature can be used:

$$\int_0^\infty e^{-x} f(x)\, dx \cong \sum_{k=1}^n A_k f(x_k) \tag{30}$$

The Gauss–Laguerre form would be appropriate, for example, for evaluation of the gamma function discussed previously.

REFERENCES

F. S. Acton, *Numerical Methods that Work*, Harper and Row, New York, 1970, Chapter 4.

P. J. Davis and I. Polonsky, "Numerical Interpolation, Differentiation, and Integration," in *Handbook of Mathematical Functions with Formulas, Graphs, and Mathematical Tables*, M. Abramowitz and I. A. Stegun (Eds.), National Bureau of Standards Applied Mathematics Series, No. 55, U.S. Government Printing Office, Washington, DC, 1964, Section 25, pp. 875–924.

A. C. Norris, *Computational Chemistry. An Introduction to Numerical Methods*, Wiley, New York, 1981, Chapter 6.

6

NUMERICAL SOLUTION OF DIFFERENTIAL EQUATIONS

6.1 FIRST-ORDER DIFFERENTIAL EQUATIONS

The need to solve ordinary, first-order differential equations,

$$y'(x) = \frac{dy}{dx} = f(x, y) \tag{1}$$

is very common in science. Solving such a differential equation means finding a solution function $y(x)$ such that it satisfies the differential equation, that is, $y'(x) = f(x, y)$, and also satisfies one initial condition, usually $y(x_0) = y_0$, where the (x_0, y_0) pair is given as part of the problem. Finding the analytical solution function $y(x)$ is always desirable if it is possible.

However, many differential equations that arise in science cannot be solved analytically. Therefore, they must be solved by numerical methods. Solving a differential equation numerically involves calculating the behaviour of the solution function one step at a time and thus building up a representation of the solution function in tabular form. The solution is obtained as a set of points that lie close to the solution function. For the first-order differential equations we are examining, these tabulated values can be plotted as $y(x)$ versus x for convenience.

Forward Tracing Methods

The general approach to the numerical solution of differential equations involves stepping in small increments in order to trace out the form of the solution. The initial point (x_0, y_0) is usually given as part of the statement of the problem. Evaluation of the differential equation at this point yields $f(x_0, y_0)$, which is the slope of the solution function at this initial point. Then one can estimate the value for the solution function at $x_0 + \Delta x$ by extrapolating (these relationships are shown in Figure 1):

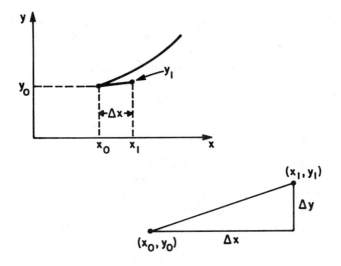

Figure 1. Plots of the quantities used by forward-tracing methods.

$$y_1 = y_0 + \Delta y = y_0 + (\Delta x)f(x_0, y_0) \tag{2}$$

At the end of the sequence, we have the new point (x_1, y_1), and the sequence can be repeated a second time to get (x_2, y_2), and so forth. After a number of steps have been taken, the solution function's tabulated values are available over an interval from (x_0, y_0) to wherever the iterative sequence was terminated.

As this method steps from one interval to the next, the true solution function and the numerical calculation diverge. The errors are cumulative in forward tracing methods.

The quality of the solution depends on the size of the Δx chosen. As Δx become smaller, the divergence between the true solution function and the numerical approximation decreases. However, the effort of calculation increases as Δx decreases because more function evaluations must be done. When working on practical problems, it is common to investigate the size of Δx necessary to achieve the precision desired.

The simple method for forward tracing described above can be explained in terms of a Taylor series expansion truncated after the inclusion of only the two leading terms:

$$y(x_0 + \Delta x) = y(x_0) + \Delta x y'(x_0) + \frac{(\Delta x)^2}{2!} y''(x_0) + \frac{(\Delta x)^3}{3!} y'''(x_0) + \cdots \tag{3}$$

One could include more than just the two leading terms, and this should lead to an improvement in the estimate of $y(x_0 + \Delta x)$. However, it would require the evaluation of higher order derivatives. Only the first derivative of the solution function $y'(x)$ is presented as part of the problem. Higher order derivatives usually are tedious to obtain for the complicated functions that arise in real problems, so this is an unattractive prospect. It can be shown that the estimates of $y(x_0 + \Delta x)$ that would be calculated using many terms from the Taylor series can be generated by a different route that is computationally superior.

Runge–Kutta Methods

About 1900, the mathematicians Runge and Kutta discovered a method for stepping across an interval from x_0 to $x_0 + \Delta x$. This can be done by computing intermediate points such that the final Δy value computed is the same as if it were derived from a preselected number of terms of the Taylor series. This is done by repeated evaluations of the differential equation itself.

The equations of the widely used fourth-order Runge–Kutta method are

$$y_{i+1} = y_i + \tfrac{1}{6} \Delta x (c_1 + 2c_2 + 2c_3 + c_4)$$

$$c_1 = f(x_i, y_i) \qquad\qquad c_2 = f(x_i + \tfrac{1}{2} \Delta x, y_i + \tfrac{1}{2} \Delta x \, c_1)$$

$$c_3 = f(x_i + \tfrac{1}{2} \Delta x, y_i + \tfrac{1}{2} \Delta x \, c_2) \qquad c_4 = f(x_i + \Delta x, y_i + \Delta x \, c_3) \qquad (4)$$

The c values are values of the slope of the solution function at the starting point of the interval, $x = x_0$; the midpoint of the interval, $x = x_0 + \tfrac{1}{2} \Delta x$; and the termination of the interval, $x = x_0 + \Delta x$.

The fourth-order Runge–Kutta method is algebraically equivalent to taking the terms of the Taylor series, Equation (3), through the $(\Delta x)^4$ term. The first neglected term is the $(\Delta x)^5$ term. The first neglected term of the Taylor series dominates the error in the overall evaluation. Therefore, the error in the estimation of Δy with the fourth-order Runge–Kutta method is inversely proportional to the fifth power of Δx. A small decrease in Δx results in an enormous increase in accuracy. This is one reason for the widespread use of the fourth-order Runge–Kutta method for numerical solution of differential equations. Another useful characteristic of this method is that it is self-starting or single step. That is, only one previous value for (x_0, y_0) is needed to calculate Δy and step across an interval. This is different from other numerical methods that need several previous points and are not self-starting. A detailed discussion of the Runge–Kutta method, its error properties, and recent reports of extensions can be found in Atkinson (1978).

Program RK4

The fourth-order Runge–Kutta method for solving first-order differential equations is simple to implement. The accompanying program RK4, which numerically solves the differential equation

$$\frac{dy}{dx} = \frac{(2x - 1)y}{x^2} + 1 \tag{5}$$

The function is expressed in the statement function. While the parameters DX and N and the initial conditions could be entered by the user, for simplicity they have been incorporated into the program itself. The output is shown for a DX value of 0.1. When the program was run with a DX value of 0.05, the value of the solution function at $x = 2.0$ was identical to six significant figures with the value shown, $y = 6.42611$.

```
      PROGRAM RK4
C....
C.... RUNGE-KUTTA METHOD FOR DIFFERENTIAL EQUATIONS
C....
      DIMENSION X(100),Y(100)
      FUNC(XX,YY)=(((2.0*XX-1.0)*YY)/XX**2)+1.0
      NOUT=1
      DX=0.1
      N=11
      X(1)=1.0
      Y(1)=2.0
      DO 10 I=1,N
      C1=FUNC(X(I),Y(I))
      C2=FUNC(X(I)+0.5*DX,Y(I)+0.5*DX*C1)
      C3=FUNC(X(I)+0.5*DX,Y(I)+0.5*DX*C2)
      C4=FUNC(X(I)+DX,Y(I)+DX*C3)
      X(I+1)=X(I)+DX
      DELY=DX*(C1+C2*2.0+C3*2.0+C4)/6.0
   10 Y(I+1)=Y(I)+DELY
      WRITE (NOUT,19)
      WRITE (NOUT,29) DX,N
      WRITE (NOUT,9)
      WRITE (NOUT,39) (I,X(I),Y(I),I=1,N)
    9 FORMAT (12X,'X',8X,'Y',/)
   19 FORMAT (' 4-TH ORDER RUNGE-KUTTA ROUTINE OUTPUT',/)
   29 FORMAT (' DIVISION SIZE',F10.5,/,' NUMBER OF POINTS',I7,/)
   39 FORMAT (' ',I5,F8.2,F12.5)
      STOP
      END
```

```
4-TH ORDER RUNGE-KUTTA ROUTINE OUTPUT

DIVISION SIZE    0.10000
NUMBER OF POINTS      11

            X         Y

    1     1.00     2.00000
    2     1.10     2.31485
    3     1.20     2.65893
    4     1.30     3.03173
    5     1.40     3.43289
    6     1.50     3.86219
    7     1.60     4.31946
    8     1.70     4.80456
    9     1.80     5.31742
   10     1.90     5.85795
   11     2.00     6.42611
```

6.2 SYSTEMS OF DIFFERENTIAL EQUATIONS

Systems of linked first-order differential equations can be solved numerically with a slightly altered Runge–Kutta approach. The equations are

$$\frac{dy_1}{dx} = f_1(x, y_1, y_2, \ldots, y_n)$$

$$\frac{dy_2}{dx} = f_2(x, y_1, y_2, \ldots, y_n) \tag{6}$$

$$\vdots$$

$$\frac{dy_n}{dx} = f_n(x, y_1, y_2, \ldots, y_n)$$

The initial conditions are all given at a common point, $x = x_0$, so the values are known for $y_1(x_0)$, $y_2(x_0)$, $\ldots$, $y_n(x_0)$. The system of equations is solved by applying the fourth-order Runge–Kutta method equations to each of the n equations in parallel at each step through the procedure.

Program RUNKUT

Systems of linked first-order differential equations can be solved simultaneously, as mentioned above. Program RUNKUT is implemented to do this using a chemical example.

Three chemical compounds that can interconvert are shown in the following chemical equation:

$$A \underset{k_{21}}{\overset{k_{12}}{\rightleftharpoons}} B \underset{k_{32}}{\overset{k_{23}}{\rightleftharpoons}} C \tag{7}$$

with the rate constants for reactions. Simple first-order differential equations express the concentrations of the three chemical compounds as a function of time, as follows:

$$\frac{d[A]}{dt} = -k_{12}[A] + k_{21}[B]$$

$$\frac{d[B]}{dt} = -(k_{23} + k_{21})[B] + k_{12}[A] + k_{32}[C] \tag{8}$$

$$\frac{d[C]}{dt} = -k_{32}[C] + k_{23}[B]$$

Each equation has a term with a positive sign for the replenishment of that compound and a negative term expressing the depletion of the compound. The concentration of species A is related only to species B; the concentration of species C is related only to species B; however, the concentration of species B is related to the concentrations of both species A and C. The initial conditions would usually be given for such a problem as the concentrations of each species at time zero, that is, $[A]$, $[B]$, and $[C]$ at $t = 0$. The problem to be solved is to find $[A]$, $[B]$, and $[C]$ versus time, that is, how the concentrations vary as a function of time. The given conditions for the problem include the form of the differential equations above as well as the initial concentrations and values for the rate constants. Program RUNKUT implements the fourth-order Runge–Kutta method for solving this problem.

The variable names used in the program are the same as in the discussion here. The three differential equations have been implemented in statement functions named FUNA, FUNB, and FUNC. Even though two of the differential equations have only two dependent variables, they have been implemented with all three dependent variables in the calling lists, namely AA, BB, and CC for convenience. Note that none of the differential equations and none of the statement functions contains time explicitly. When executing, the program first prompts the user for parameter values. The values entered are echoed as a double check for accuracy of input. The actual solution of the system of equations is generated in the loop after the parameters are input. A double-loop structure is used so that it is convenient to save the values for each concentration at each 1-sec interval for later plotting. The equations within the loop are the same as in the earlier, simple Runge–Kutta program, but they have been tripled since there are three equations. After the solutions have been calculated, RUNKUT calls a plotting routine named PLOT that generates a simple character plot of the results.

Three example runs were executed and are shown. In all three, the values for the rate constants were input as $k_{12} = 1.0$, $k_{21} = 0.5$, $k_{23} = 0.5$, and $k_{32} = 0.25$. In the first run, the values for the starting concentrations were input as $[A] = 1.00$, $[B] = 0$, and $[C] = 0$. In the second run, the initial concentrations were input as $[A] = 0$, $[B] = 1.0$, and $[C] = 0$. In the third run, the initial concentrations were input as $[A] = 0$, $[B] = 0$, and $[C] = 1.0$. The three plots show how the concentrations of the three species change rapidly at first and then stabilize, reaching equilibrium after about 10 sec. Of course, the equilibrium concentrations of the three species are equal in the three runs because the direction from which equilibrium is reached is immaterial. The plotting routine is scaling the vertical axis of its plots to make the largest y value just fill the entire space allotted, since all three of these plots contain a concentration of 1.0, the scaling of the three plots is identical. Program RUNKUT could be used to investigate the effect of altering the rate constants as well as, or in addition to, the starting concentrations.

This same approach can be used to model much more complex systems of reactions. Complicated interrelationships between species, even including loops, can be modeled and the concentrations of species involved plotted as an aid to understanding the chemistry.

```
        PROGRAM RUNKUT
C....
C.... PROGRAM TO SOLVE LINKED FIRST-ORDER DIFFERENTIAL EQUATIONS
C....   USING THE 4-TH ORDER RUNGE-KUTTA METHOD
C....
        DIMENSION Y(123)
        REAL K12,K21,K23,K32
C....
C.... THE DIFFERENTIAL EQUATIONS
C....
        FUNA(AA,BB,CC) = -K12*AA + K21*BB
        FUNB(AA,BB,CC) = -(K23+K21)*BB + K12*AA + K32*CC
        FUNC(AA,BB,CC) = -K32*CC + K23*BB
        DATA NUM/41/,NOUT/1/,NPTS/123/
C....
C.... INPUT CONDITIONS
C....
        WRITE (NOUT,1)
      1 FORMAT (' THIS IS RUNKUT'/)
        WRITE (NOUT,2)
      2 FORMAT (' ENTER STARTING CONCENTRATIONS')
        READ (NOUT,*) A,B,C
        WRITE (NOUT,3) A,B,C
      3 FORMAT (' STARTING CONCENTRATIONS',/,' ',3F7.2)
        WRITE (NOUT,4)
      4 FORMAT (' ENTER NUMBER OF STEPS PER SECOND')
        READ (NOUT,*) NN
        WRITE (NOUT,5) NN
        NH=NN/2
      5 FORMAT (' NUMBER OF STEPS PER SECOND: ',I4)
        DX = 1.0/FLOAT(NN)
        WRITE (NOUT,6) DX
      6 FORMAT (' STEP SIZE: ',F7.3)
        WRITE (NOUT,7)
      7 FORMAT (' ENTER RATE CONSTANTS: K12,K21,K23,K32')
        READ (NOUT,*) K12,K21,K23,K32
        WRITE (NOUT,8) K12,K21,K23,K32
      8 FORMAT (' RATE CONSTANTS: ',/,5X,4F7.2)
        WRITE (NOUT,9)
      9 FORMAT (//,5X,'TIME',5X,'[A]',7X,'[B]',7X,'[C]',/)
        T=0.0
        WRITE (NOUT,10) T,A,B,C
     10 FORMAT (' ',F7.0,3F10.4)
        I=1
        Y(I)=A
        Y(I+NUM)=B
        Y(I+2*NUM)=C
C....
C.... SOLVE SET OF DIFFERENTIAL EQUATIONS
C....
        DO 50 J=1,20
          DO 20 K=1,NN
          C1A = FUNA(A,B,C)
          C1B = FUNB(A,B,C)
          C1C = FUNC(A,B,C)
          C2A = FUNA(A+0.5*C1A,B+0.5*C1B,C+0.5*C1C)
          C2B = FUNB(A+0.5*C1A,B+0.5*C1B,C+0.5*C1C)
          C2C = FUNC(A+0.5*C1A,B+0.5*C1B,C+0.5*C1C)
          C3A = FUNA(A+0.5*C2A,B+0.5*C2B,C+0.5*C2C)
          C3B = FUNB(A+0.5*C2A,B+0.5*C2B,C+0.5*C2C)
          C3C = FUNC(A+0.5*C2A,B+0.5*C2B,C+0.5*C2C)
          C4A = FUNA(A+C3A,B+C3B,C+C3C)
          C4B = FUNB(A+C3A,B+C3B,C+C3C)
          C4C = FUNC(A+C3A,B+C3B,C+C3C)
```

```
          DELA = (C1A+2.0*C2A+2.0*C3A+C4A)*DX/6.0
          DELB = (C1B+2.0*C2B+2.0*C3B+C4B)*DX/6.0
          DELC = (C1C+2.0*C2C+2.0*C3C+C4C)*DX/6.0
          T = T + DX
          A = A + DELA
          B = B + DELB
          C = C + DELC
          IF (K.NE.NH) GO TO 20
          I=I+1
          Y(I)=A
          Y(I+NUM)=B
          Y(I+2*NUM)=C
   20     CONTINUE
          WRITE (NOUT,10) T,A,B,C
          I=I+1
          Y(I)=A
          Y(I+NUM)=B
          Y(I+2*NUM)=C
   50     CONTINUE
C....
C.... SET PARAMETERS AND CALL PLOTTING ROUTINE
C....
          YMAX=0.0
          DO 60 I=1,NPTS
          IF (Y(I).GT.YMAX) YMAX=Y(I)
   60     CONTINUE
          YMIN=0.0
          NGR=3
          CALL PLOT (Y,NPTS,YMIN,YMAX,NGR)
          STOP
          END
C------------------------------------------------------------
          SUBROUTINE PLOT (Y,NDIM,YMIN,YMAX,NGR)
C....
C.... ROUTINE TO GENERATE CHARACTER PLOT OF UP TO THREE FUNCTIONS
C....   FIRST FUNCTION REPRESENTED BY A, SECOND BY B, THIRD BY C
C....
          DIMENSION Y(NDIM),A(79),NZ(3),CHAR(3)
          DATA BLANK,PLUS,CHAR(1),CHAR(2),CHAR(3)/' ','+','A','B','C'/
          DATA NW/79/,NOUT/1/
          NPTS=NDIM/NGR
          DELTA=(YMAX-YMIN)/70.0
          YMN=YMIN-4.0*DELTA
          DO 10 I=1,NW
   10     A(I)=BLANK
          DO 60 I=1,NPTS
          IF (MOD(I,6)-1) 24,17,24
   17     DO 20 J=5,NW,10
   20     A(J)=PLUS
   24     DO 30 J=1,NGR
          N=(Y(I+(J-1)*NPTS)-YMN)/DELTA+1.5
          IF (N.LT.1) N=1
          IF (N.GT.NW) N=NW
          NZ(J)=N
   30     A(N)=CHAR(J)
          WRITE (NOUT,19) A
          DO 40 J=1,NGR
   40     A(NZ(J))=BLANK
          IF (MOD(I,6)-1) 60,38,60
   38     DO 50 J=5,NW,10
   50     A(J)=BLANK
   60     CONTINUE
   19     FORMAT (' ',79A1)
          RETURN
          END
```

```
THIS IS RUNKUT

ENTER STARTING CONCENTRATIONS
1.  0.  0.
 STARTING CONCENTRATIONS
     1.00    0.00    0.00
 ENTER NUMBER OF STEPS PER SECOND
25
 NUMBER OF STEPS PER SECOND:   25
 STEP SIZE:  0.040
 ENTER RATE CONSTANTS:  K12,K21,K23,K32
1.  0.5 0.5 0.25
 RATE CONSTANTS:
           1.00    0.50    0.50    0.25
          TIME     [A]      [B]      [C]
           0.    1.0000   0.0000   0.0000
           1.    0.6319   0.2250   0.1432
           2.    0.4297   0.3081   0.2622
           3.    0.3155   0.3312   0.3533
           4.    0.2493   0.3309   0.4198
           5.    0.2098   0.3232   0.4671
           6.    0.1856   0.3143   0.5001
           7.    0.1705   0.3066   0.5229
           8.    0.1610   0.3005   0.5385
           9.    0.1548   0.2961   0.5491
          10.    0.1508   0.2929   0.5564
          11.    0.1481   0.2906   0.5613
          12.    0.1464   0.2890   0.5646
          13.    0.1452   0.2880   0.5668
          14.    0.1444   0.2872   0.5683
          15.    0.1439   0.2867   0.5693
          16.    0.1436   0.2864   0.5700
          17.    0.1433   0.2862   0.5705
          18.    0.1432   0.2860   0.5708
          19.    0.1431   0.2859   0.5710
          20.    0.1430   0.2858   0.5711
```

```
THIS IS RUNKUT

ENTER STARTING CONCENTRATIONS
0. 1. 0.
 STARTING CONCENTRATIONS
    0.00    1.00    0.00
 ENTER NUMBER OF STEPS PER SECOND
25
 NUMBER OF STEPS PER SECOND:   25
 STEP SIZE:   0.040
 ENTER RATE CONSTANTS:  K12,K21,K23,K32
1. 0.5 0.5 0.25
 RATE CONSTANTS:
        1.00    0.50    0.50    0.25


   TIME      [A]         [B]         [C]

    0.      0.0000      1.0000      0.0000
    1.      0.1125      0.6677      0.2199
    2.      0.1540      0.4952      0.3507
    3.      0.1656      0.4039      0.4305
    4.      0.1655      0.3542      0.4803
    5.      0.1616      0.3265      0.5119
    6.      0.1572      0.3106      0.5322
    7.      0.1533      0.3013      0.5454
    8.      0.1503      0.2956      0.5541
    9.      0.1480      0.2921      0.5599
   10.      0.1464      0.2898      0.5637
   11.      0.1453      0.2884      0.5662
   12.      0.1445      0.2875      0.5679
   13.      0.1440      0.2869      0.5691
   14.      0.1436      0.2865      0.5698
   15.      0.1434      0.2862      0.5704
   16.      0.1432      0.2861      0.5707
   17.      0.1431      0.2859      0.5709
   18.      0.1430      0.2859      0.5711
   19.      0.1430      0.2858      0.5712
   20.      0.1429      0.2858      0.5713
```

```
THIS IS RUNKUT

ENTER STARTING CONCENTRATIONS
0.  0.  1.
STARTING CONCENTRATIONS
     0.00    0.00    1.00
ENTER NUMBER OF STEPS PER SECOND
25
NUMBER OF STEPS PER SECOND:   25
STEP SIZE:   0.040
ENTER RATE CONSTANTS: K12, K21, K23, K32
1.0 0.5 0.5 0.25
RATE CONSTANTS:
          1.00    0.50    0.50    0.25
```

TIME	[A]	[B]	[C]
0.	0.0000	0.0000	1.0000
1.	0.0358	0.1099	0.8543
2.	0.0656	0.1754	0.7591
3.	0.0883	0.2153	0.6964
4.	0.1049	0.2402	0.6549
5.	0.1168	0.2559	0.6273
6.	0.1250	0.2661	0.6089
7.	0.1307	0.2727	0.5965
8.	0.1346	0.2771	0.5883
9.	0.1373	0.2799	0.5828
10.	0.1391	0.2819	0.5790
11.	0.1403	0.2831	0.5765
12.	0.1411	0.2840	0.5749
13.	0.1417	0.2845	0.5737
14.	0.1421	0.2849	0.5730
15.	0.1423	0.2852	0.5725
16.	0.1425	0.2854	0.5721
17.	0.1426	0.2855	0.5719
18.	0.1427	0.2855	0.5717
19.	0.1427	0.2856	0.5716
20.	0.1428	0.2856	0.5715

6.3 PREDICTOR–CORRECTOR METHODS

An alternative method to step across an interval Δx to trace out the solution function of a first-order differential equation involves using more than one previous point during the calculations. These methods are called multistep methods, and they have been developed in a number of formulations. An excellent discussion of predictor–corrector methods can be found in Acton (1970).

Typically, multistep methods extrapolate from the point $y_0 = y(x_0)$ to the new point $y_1 = y(x_1)$ using an approximate predictor step followed by a corrector step that improves the approximation. The predictor step is done by fitting a polynomial to y_0' and the two previous points, y_{-1}' and y_{-2}', as seen in Figure 2. The three points uniquely define a second-order polynomial, a parabola. Then the fit parabola is extrapolated across the interval Δx. The area under the parabola can be calculated exactly from any of the previous points y_k'. Then this area under the dy/dx curve is added to the appropriate y value to get a predicted value for y_1 called y_{1p}. In Figure 2, the area was computed only between x_0 and x_1 and then this area was added to y_0 to get y_{1p}. The quality of this extrapolation will be only as good as the degree to which a parabola mimics the differential equation and the size of Δx.

The differential equation itself can be used to improve on this first predicted value of y_{1p}. The y_{1p} value, along with x_1, can be put into the differential equation to get a predicted value of the derivative at x_1, that is, $y_1' = f(x_1, y_{1p})$. This estimate of y_1' can be used along with the two previous points, y_0' and y_{-1}', to fit a new, better parabola that can be used to calculate a new, better area that can be used to calculate a new, better y_1, now called y_{1c}. This is the corrected value.

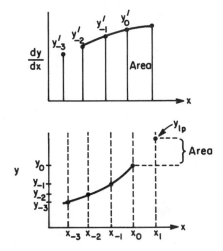

Figure 2. Plots of the quantities used by the Milne predictor–corrector method.

The sequence of calculations for the Milne method is as follows, given values for y_0, y_{-1}, y_{-2}, y_{-3}, y_0', y_{-1}', y_{-2}', y_{-3}', and $dy/dx = f(x, y)$:

$$y_{1p} = y_{-3} + \tfrac{1}{3}[4(\Delta x)](2y_{-2}' - y_{-1}' + 2y_0')$$

$$y_1' = f(x_1, y_{1p})$$

$$y_{1c} = y_{-1} + \tfrac{1}{3}\Delta x(y_{-1}' + 4y_0' + y_1')$$

$$y_1' = f(x_1, y_{1c})$$

(9)

An important feature of the predictor–corrector method is that the method provides a method for following the accuracy of the numerical evaluation as part of the calculation sequence. The difference between y_{1p} and y_{1c} provides information on whether Δx is too large. If Δx is too large, it can be halved, which will markedly increase the accuracy of the overall method since the Milne method has an error that is inversely proportional to the fifth power of Δx. A more detailed discussion of the error properties of the predictor–corrector methods can be found in Acton (1970) or Ralstron and Wilf (1960).

REFERENCES

F. S. Acton, *Numerical Methods that Work*, Harper & Row, New York, 1970.

K. E. Atkinson, *An Introduction to Numerical Analysis*, Wiley, New York, 1978, pp. 366ff.

B. Carnahan, H. A. Luther, and J. O. Wilkes, *Applied Numerical Methods*, Wiley, New York, 1969.

A. Ralston and H. S. Wilf, *Mathematical Methods for Digital Computers*, Wiley, New York, 1960.

7

MATRIX METHODS
AND SYSTEMS OF
LINEAR EQUATIONS

7.1 METHODS FOR SOLVING SYSTEMS OF LINEAR EQUATIONS

Many chemistry problems require the solution of a set of linear equations. Linear equations are those in which the quantities to be determined appear in a linear relationship. Examples of places in chemistry where systems of simultaneous equations must be solved include mixture analysis in spectrophotometry, multicomponent equilibrium systems, and least-squares analysis.

The methods we will cover in this section can be described most easily using the notation and ideas of matrix algebra. The reader is assumed to be familiar with basic matrix algebra. Many linear algebra texts cover the topic in great detail, and one of them could be consulted for review (e.g., Hohn 1964, Boas 1983). The following discussion introduces the notation to be used in this section.

A matrix, represented by the symbol **A**, is

$$\mathbf{A} = (a_{ij}) = \begin{bmatrix} a_{11} & a_{12} & a_{13} & \cdots & a_{1n} \\ a_{21} & a_{22} & a_{23} & \cdots & a_{2n} \\ \vdots & \vdots & \vdots & \ddots & \vdots \\ a_{m1} & a_{m2} & a_{m3} & \cdots & a_{mn} \end{bmatrix} \tag{1}$$

This matrix has m rows and n columns and is said to be an $m \times n$ matrix. Each lowercase a is an element of the matrix, and the subscripts denote the row and column number of the element.

Many elementary algebraic operations are defined for matrices. Two matrices can be added only if they have the same dimensions, the same number of rows and columns. Addition of two matrices **A** and **B** to give **C** is performed by adding corresponding pairs of elements drawn from the two matrices. In matrix notation, addition is $\mathbf{C} = \mathbf{A} + \mathbf{B}$. In algebraic notation, we write $c_{ij} = a_{ij} + b_{ij}$ for all i and all j. Matrix addition obeys the commutative law, $\mathbf{A} + \mathbf{B} = \mathbf{B} + \mathbf{A}$, and the associative law, $(\mathbf{A} + \mathbf{B}) + \mathbf{C} = \mathbf{A} + (\mathbf{B} + \mathbf{C})$.

Multiplication of a matrix by a constant is done by multiplying each element in the matrix by the constant. Two matrices can be multiplied in order $\mathbf{AB}$ only if the number of columns in $\mathbf{A}$ equals the number of rows in $\mathbf{B}$. The product $\mathbf{AB}$ of two matrices is also a matrix $\mathbf{C}$. The (i, j)th element of the product matrix is the sum of the products by pairs of the elements of the ith row of $\mathbf{A}$ and the jth column of $\mathbf{B}$. In algebraic notation,

$$c_{ij} = \sum_{k=1}^{n} a_{ik} b_{kj} \quad \text{for all } i, j \tag{2}$$

The product matrix $\mathbf{C}$ has the same number of rows as $\mathbf{A}$ and columns as $\mathbf{B}$. Matrix multipliction does not obey the commutative law; therefore, $\mathbf{AB}$ does not necessarily equal $\mathbf{BA}$. Matrix multiplication does obey the associative law, so $\mathbf{A}(\mathbf{BC}) = \mathbf{AB}(\mathbf{C})$.

There is no direct matrix algebra equivalent of division, although the matrix operation called inversion is, in some ways, analogous to division. Inversion is only defined for square matrices, those for which the number of rows equals the number of columns. Then the inverse of the square matrix $\mathbf{A}$ is $\mathbf{B}$ where these two matrices obey the equality $\mathbf{AB} = \mathbf{I} = \mathbf{BA}$. The matrix $\mathbf{I}$ is a special matrix named the identity matrix, which has all the elements along its main diagonal equal to 1, $a_{ii} = 1$, and all other elements equal to 0. The inverse matrix of a matrix $\mathbf{a}$ is given the special notation $\mathbf{A}^{-1}$ so

$$\mathbf{AA}^{-1} = \mathbf{A}^{-1}\mathbf{A} = \mathbf{I} \tag{3}$$

Finding the elements of the inverse matrix for a given matrix is one of the points of this section, since this operation is equivalent to solving a set of linear equations.

Systems of Linear Equations

The usual notation for a system of linear equations is

$$\begin{aligned}
a_{11}x_1 + a_{12}x_2 + \cdots + a_{1n}x_n &= b_1 \\
a_{21}x_1 + a_{22}x_2 + \cdots + a_{2n}x_n &= b_2 \\
&\vdots \\
a_{m1}x_m + a_{m2}x_m + \cdots + a_{mn}x_n &= b_m
\end{aligned} \tag{4}$$

Due to the definitions of matrix algebra, this system of linear equations can be written as

$$\mathbf{AX} = \mathbf{B} \tag{5}$$

where $\mathbf{A}$ is the $n \times m$ coefficient matrix, $\mathbf{X}$ is the $n \times 1$ solution vector being sought, and $\mathbf{B}$ is the $n \times 1$ constant vector. In order to be able to solve the system of linear equations for the x values, we must have a square coefficient matrix. That is, there must be as many equations in the set as there are unknowns per equation.

Cramer's Rule

Systems of simultaneous linear equations can be solved by employing Cramer's rule. For a system of $n \times n$ equations, the value of the ith unknown is given by the ratio of two determinants. The denominator is the determinant of the coefficient matrix; the numerator is the same determinant except that the ith column is replaced by the column of constants. For the 2×2 set of equations

$$a_{11}x_1 + a_{12}x_2 = b_1$$

$$a_{21}x_1 + a_{22}x_2 = b_2$$

(6)

the solutions are given by

$$x_1 = \frac{1}{D} \begin{vmatrix} b_1 & a_{12} \\ b_2 & a_{22} \end{vmatrix} \qquad x_2 = \frac{1}{D} \begin{vmatrix} a_{11} & b_1 \\ a_{21} & b_2 \end{vmatrix} \qquad D = \begin{vmatrix} a_{11} & a_{12} \\ a_{21} & a_{22} \end{vmatrix} \qquad (7)$$

By definition, 2×2 determinants are evaluated as follows:

$$\begin{vmatrix} a_{11} & a_{12} \\ a_{21} & a_{22} \end{vmatrix} = a_{11}a_{22} - a_{21}a_{12} \qquad (8)$$

This requires two multiplications and one subtraction. Thus, the two equations for x_1 and x_2 above can be easily evaluated. As the size of the determinant grows, the effort of evaluation grows too. For a 3×3 determinant, evaluation is as follows:

$$\begin{vmatrix} a_{11} & a_{12} & a_{13} \\ a_{21} & a_{22} & a_{23} \\ a_{31} & a_{32} & a_{33} \end{vmatrix} = \begin{aligned} & a_{11}a_{22}a_{33} + a_{12}a_{23}a_{31} + a_{13}a_{21}a_{32} - \\ & a_{31}a_{22}a_{13} - a_{32}a_{23}a_{11} - a_{33}a_{21}a_{12} \end{aligned} \qquad (9)$$

This expression has $n! = 6$ terms, each one of which is the product of n numbers, so $(n)(n!) = 18$ operations are needed to evaluate it. Determinants of larger size can be evaluated by reduction using cofactors to a series of 3×3 determinants. Algorithms can be devised that allow evaluation of $n \times n$ determinants with approximately $2n!$ multiplications. This factor grows so rapidly as a function of n that this approach for the evaluation of determinants is impractical. Evaluation of one 20×20 determinant would require on the order of 10^{18} operations that would take billions of years.

There are alternative ways to evaluate determinants that are much more efficient than the brute force method described above. Elementary row operations can be used to make a determinant triangular, and then it can be evaluated easily. This sequence of operations can be implemented in an algorithm where the number of multiplications goes as $n^3/3$. To evaluate a 20×20 determinant by this approach requires only a few thousand operations. However, even so, Cramer's rule is not used because superior, more efficient alternatives exist.

The Gauss–Seidel Method

Systems of linear equations can be solved using an iterative approach that attempts to converge to a solution by making incremental improvements in the x values. One such method is known as the Gauss–Seidel method. Consider a 3×3 example:

$$
\begin{aligned}
a_{11}x_1 + a_{12}x_2 + a_{13}x_3 &= b_1 \\
a_{21}x_1 + a_{22}x_2 + a_{23}x_3 &= b_2 \\
a_{31}x_1 + a_{32}x_2 + a_{33}x_3 &= b_3
\end{aligned}
\tag{10}
$$

These equations can be rearranged to the form

$$
\begin{aligned}
x_1 &= 1/a_{11}(b_1 - a_{12}x_2 - a_{13}x_3) \\
x_2 &= 1/a_{22}(b_2 - a_{21}x_1 - a_{23}x_3) \\
x_3 &= 1/a_{33}(b_3 - a_{31}x_1 - a_{32}x_2)
\end{aligned}
\tag{11}
$$

To use this method, estimate initial values for x_2 and x_3 and plug them into the equation for x_1. The value for x_1 and the initial estimate for x_3 are plugged into the equation for x_2. The values for x_1 and x_2 are plugged into the third equation. The three steps are repeated again and again, until sufficient accuracy is obtained. Each trip through the cycle takes n^2 multiplications.

In general, for an $n \times n$ set of equations, the mth approximation for the ith variable is given by

$$
x_i^{(m)} = \frac{1}{a_{ii}} \left[b_i - \sum_{j=1}^{i-1} a_{ij}x_j^{(m)} - \sum_{j=i+1}^{n} a_{ij}x_j^{(m-1)} \right]
\tag{12}
$$

A drawback to this method is that it is not guaranteed to converge to the correct solution. However, if the coefficient matrix is diagonally dominant, then the method is guaranteed to converge. A diagonally dominant matrix is one for which

$$
|a_{ii}| \geq \sum_{\substack{j=1 \\ j \neq i}}^{n} |a_{ij}|
\tag{13}
$$

That is, the diagonal element in each row of the matrix must be greater than the sum of the absolute values of all other elements in that row. In some cases, diagonal dominance can be achieved by rearranging the set of equations to place the largest elements on the main diagonal.

Gauss–Jordan Elimination Method

Systems of linear equations can be solved by a direct method that systematically converts the given coefficient matrix into a form that is more easily solved. The conversion is accomplished by repeated use of elementary row operations to the coefficient matrix.

To understand the basis of the Gauss–Jordon method, it is necessary to introduce some additional matrix algebra. Recall the definition and properties of the inverse of a matrix described earlier in this section. No mention was made of how to find the inverse of a given matrix. An effective way to perform this operation is as follows. Starting with the coefficient $\mathbf{A}$ and the identity matrix $\mathbf{I}$, perform elementary row operations on $\mathbf{A}$ to transform $\mathbf{A}$ to $\mathbf{I}$. Do each step to $\mathbf{I}$ as well, and when finished, $\mathbf{A}$ has become $\mathbf{I}$ and $\mathbf{I}$ has become $\mathbf{A}^{-1}$. The elementary row operations are:

1. Interchange any two rows.
2. Multiply the elements of a row by a nonzero constant.
3. Add a multiple of any row to another row.

Once $\mathbf{A}^{-1}$ is obtained by this method, then the values of the unknowns can be obtained by the simple matrix multiplication operation

$$\mathbf{X} = \mathbf{A}^{-1}\mathbf{B} \tag{14}$$

It turns out that all the necessary operations can be carried out at once, and the method implementing this is known as the Gauss–Jordan method. This method involves the following steps. First, an augmented matrix $(\mathbf{A} \mid \mathbf{B})$ is formed:

$$(\mathbf{A} \mid \mathbf{B}) = \begin{bmatrix} a_{11} & a_{12} & \cdots & a_{1n} & b_1 \\ a_{21} & a_{22} & \cdots & a_{2n} & b_2 \\ \vdots & \vdots & \ddots & \vdots & \vdots \\ a_{n1} & a_{n2} & \cdots & a_{nn} & b_n \end{bmatrix} \tag{15}$$

Then elementary row operations are used to create the matrix

$$\begin{bmatrix} f_1 & 0 & \cdots & 0 & g_1 \\ 0 & f_2 & \cdots & 0 & g_2 \\ \vdots & \vdots & \ddots & \vdots & \vdots \\ 0 & 0 & \cdots & f_n & g_n \end{bmatrix} \tag{16}$$

Then the solutions to the system of equations are just

$$x_i = \frac{g_i}{f_i} \quad \text{for all } i \tag{17}$$

The step-by-step solution of the following 3×3 set of equations shows how the Gauss–Jordan method works in practice:

$$\begin{bmatrix} 5x_1 + 2x_2 + x_3 = 36 \\ x_1 + 7x_2 + 3x_3 = 63 \\ 2x_1 + 3x_2 + 8x_3 = 81 \end{bmatrix} \tag{18}$$

The augmented matrix to be worked on is

$$\begin{bmatrix} 5 & 2 & 1 & 36 \\ 1 & 7 & 3 & 63 \\ 2 & 3 & 8 & 81 \end{bmatrix} \tag{19}$$

Our objective is to use elementary row operations to transform this matrix to the form

$$\begin{bmatrix} 1 & 0 & 0 & x_1 \\ 0 & 1 & 0 & x_2 \\ 0 & 0 & 1 & x_3 \end{bmatrix} \tag{20}$$

The first operation, applied to the augmented matrix, is to divide row 1 by the element $a_{11} = 5$ to get

$$\begin{bmatrix} 1 & 0.4 & 0.2 & 7.2 \\ 1 & 7 & 3 & 63 \\ 2 & 3 & 8 & 81 \end{bmatrix} \tag{21}$$

This puts a 1 in position $(1, 1)$. Then we multiply row 1 by $a_{21} = 1$ and subtract from row 2 to get

$$\begin{bmatrix} 1 & 0.4 & 0.2 & 7.2 \\ 0 & 6.6 & 2.8 & 55.8 \\ 2 & 3 & 8 & 81 \end{bmatrix} \tag{22}$$

This puts a zero in position $(2, 1)$. Then we multiply row 1 by $a_{31} = 2$ and subtract from row 3 to get

$$\begin{bmatrix} 1 & 0.4 & 0.2 & 7.2 \\ 0 & 6.6 & 2.8 & 55.8 \\ 0 & 2.2 & 7.6 & 66.6 \end{bmatrix} \tag{23}$$

Now all the elements below the main diagonal in column 1 are zero. Now we will transform the second column. We start by dividing all the elements of row 2 by $a_{22} = 6.6$ to get

$$\begin{bmatrix} 1 & 0.4 & 0.2 & 7.2 \\ 0 & 1 & 0.425 & 8.46 \\ 0 & 2.2 & 7.6 & 66.6 \end{bmatrix} \tag{24}$$

which makes $a_{22} = 1$. We then multiply row 2 by $a_{12} = 0.4$ and subtract from row 1. We multiply row 2 by $a_{32} = 2.2$ and subtract from row 3 to get

$$\begin{bmatrix} 1 & 0 & 0.03 & 3.82 \\ 0 & 1 & 0.425 & 8.46 \\ 0 & 0 & 6.68 & 47.99 \end{bmatrix} \tag{25}$$

The second column is finished. We divide row 3 by $a_{33} = 6.68$ to make $a_{33} = 1$. We multiply row 3 by $a_{23} = 0.425$ and subtract from row 2; we multiply row 3 by $a_{13} = 0.03$ and subtract from row 1 to get

$$\begin{bmatrix} 1 & 0 & 0 & 3.6 \\ 0 & 1 & 0 & 5.4 \\ 0 & 0 & 1 & 7.2 \end{bmatrix} \tag{26}$$

Therefore, $x_1 = 3.6$, $x_2 = 5.4$, and $x_3 = 7.2$.

The Gauss–Jordan method can be implemented with a quite short routine. The general procedure is

(Row k)/a_{kk}

(Row 1) − (a_{1k})(Row k)

(Row 2) − (a_{2k})(Row k)

$\quad\vdots$ do for all k, $k = 1, 2, \ldots, n$

(exclude row k itself)

$\quad\vdots$

(Row n) − (a_{nk})(Row k)

This simplest procedure assumes that no diagonal element will equal zero. The accompanying subroutine, named GJE, presents this simplest algorithm coded in FORTRAN. It does not include pivoting, so it could attempt to divide by zero and therefore abort if a singular matrix were supplied to it. Therefore, GJE is for demonstration purposes only.

GAUSS-JORDAN ELIMINATION ROUTINE

```
      SUBROUTINE GJE (A,X,N)
C.... GAUSS-JORDAN ELIMINATION METHOD
C.... REDUCE COEFFICIENT MATRIX TO DIAGONAL FORM
      DOUBLE PRECISION A(N,N),X(N),MULT
      DO 3 I=1,N
      DO 2 J=1,N
      IF (I.EQ.J) GO TO 2
      MULT=A(J,I)/A(I,I)
      X(J)=X(J)-X(I)*MULT
      II=I+1
      DO 1 K=II,N
      A(J,K)=A(J,K)-MULT*A(I,K)
    1 CONTINUE
    2 CONTINUE
    3 CONTINUE
C.... CALCULATE THE SOLUTION VECTOR
      DO 4 I=1,N
    4 X(I)=X(I)/A(I,I)
      RETURN
      END
```

7.2 IMPLEMENTATION OF THE GAUSS–JORDAN METHOD

Program LINEQ and Subroutine DLIN

The subroutine DLIN implements the Gauss–Jordan elimination method with complete pivoting to perform three simultaneous operations on a system of equations in matrix form. The three operations performed are to calculate the value of the determinant of the matrix, to calculate the inverse of the matrix, and to solve for the values of the unknowns. The arguments in the calling list to DLIN are as follows: A is the coefficient matrix; B is the constant vector; N is the order of A and the number of unknowns; and DET is the variable in which the value of the determinant is returned to the calling program. The matrix A is returned as the inverse of the original coefficient matrix A, and B is returned as the solution vector. In order to achieve maximum precision in the computation, all real variables in DLIN are declared to be double precision, so they must be declared double precision in the calling routine too.

Program LINEQ is a short, simple main routine that obtains values for the coefficient matrix and constant vector from the user and calls DLIN to solve the system of linear equations. The input is interactive and interrogative; the user is asked a series of questions, and the answers define the problem being solved.

As shown, LINEQ and DLIN have been run with two different sets of values for the coefficient values in the matrix A and constant values B.

```
      PROGRAM LINEQ
C....
C.... SOLUTION OF SET OF LINEAR EQUATIONS USING THE GAUSS-JORDAN
C....    ELIMINATION METHOD
C....
      DOUBLE PRECISION A(10,10),B(10),DET
      DATA NOUT/1/
C....
C.... INPUT PARAMETERS AND PRINT OUT PROBLEM
C....
      WRITE (NOUT,101)
  101 FORMAT (' LINEAR EQUATIONS SOLUTION ROUTINE',/)
      WRITE (NOUT,102)
  102 FORMAT (' INPUT THE NUMBER OF EQUATIONS IN THE SET')
      READ (NOUT,*) N
      WRITE (NOUT,103) N
  103 FORMAT (' INPUT THE',I3,' CONSTANTS')
      READ (NOUT,*) (B(I),I=1,N)
      DO 110 II=1,N
      WRITE (NOUT,104) N,II
  104 FORMAT (' INPUT THE',I3,' COEFFICIENTS FOR EQUATION',I3)
      READ (NOUT,*) (A(II,J),J=1,N)
  110 CONTINUE
      WRITE (NOUT,111)
  111 FORMAT (//,' THE SET OF EQUATIONS: ',/)
      DO 120 I=1,N
      WRITE (NOUT,112) (A(I,J),J=1,N),B(I)
  112 FORMAT (' ',6F10.2)
  120 CONTINUE
C....
C.... CALL DLIN TO SOLVE EQUATIONS
C....
      CALL DLIN (A,B,N,DET)
C....
C.... PRINT OUT RESULTS
C....
      WRITE (NOUT,201)
  201 FORMAT (/,' INVERSE MATRIX',/)
      DO 210 I=1,N
      WRITE (NOUT,202) (A(I,J),J=1,N)
  202 FORMAT (' ',6F10.5)
  210 CONTINUE
      WRITE (NOUT,211) (B(I),I=1,N)
  211 FORMAT (/,' SOLUTIONS: ',/,(6F10.5))
      WRITE (NOUT,212) DET
  212 FORMAT (/,' DETERMINANT VALUE: ',E15.5)
      STOP
      END
C------------------------------------------------------------------
      SUBROUTINE DLIN (A,B,N,DET)
      IMPLICIT REAL*8 (A-H,O-Z)
      DIMENSION A(10,10),B(10)
      DIMENSION ISWTCH(10),JROW(10),JCOL(10)
      DET=1.D0
      DO 15 J=1,N
   15 ISWTCH(J)=0
      DO 215 I=1,N
      PIVMAX=0.
      DO 65 J=1,N
      IF (ISWTCH(J))   25,25,65
   25 DO 55 K=1,N
      IF (ISWTCH(K))   35,35,55
   35 IF ( DABS(PIVMAX)- DABS(A(J,K)))   45,55,55
   45 IROW=J
      ICOL=K
      PIVMAX=A(J,K)
   55 CONTINUE
   65 CONTINUE
      ISWTCH(ICOL)=1
      IF (IROW-ICOL)   75,115,75
```

```
 75 DET=-DET
    DO 85 L=1,N
    SAVE=A(IROW,L)
    A(IROW,L)=A(ICOL,L)
 85 A(ICOL,L)=SAVE
    SAVE=B(IROW)
    B(IROW)=B(ICOL)
    B(ICOL)=SAVE
115 JROW(I)=IROW
    JCOL(I)=ICOL
    PIVEL=A(ICOL,ICOL)
    DET=DET*PIVEL
    A(ICOL,ICOL)=1.0
    DO 135 L=1,N
135 A(ICOL,L)=A(ICOL,L)/PIVEL
    B(ICOL)=B(ICOL)/PIVEL
    DO 215 J=1,N
    IF (J-ICOL)  175,215,175
175 PIVMAX=A(J,ICOL)
    A(J,ICOL)=0.
    DO 185 L=1,N
185 A(J,L)=A(J,L)-A(ICOL,L)*PIVMAX
205 B(J)=B(J)-B(ICOL)*PIVMAX
215 CONTINUE
    RETURN
    END
```

```
         LINEAR EQUATIONS SOLUTION ROUTINE

         INPUT THE NUMBER OF EQUATIONS IN THE SET
    4
         INPUT THE  4 CONSTANTS
    0.102 0.100 0.219 0.035
         INPUT THE  4 COEFFICIENTS FOR EQUATION  1
    1.502 0.051 0.0 0.041
         INPUT THE  4 COEFFICIENTS FOR EQUATION  2
    0.026 1.152 0.0 0.082
         INPUT THE  4 COEFFICIENTS FOR EQUATION  3
    0.034 0.036 2.532 0.293
         INPUT THE  4 COEFFICIENTS FOR EQUATION  4
    0.034 0.068 0.0 0.347


    THE SET OF EQUATIONS:

         1.50       0.05       0.00       0.04       0.10
         0.03       1.15       0.00       0.08       0.10
         0.03       0.04       2.53       0.29       0.22
         0.03       0.07       0.00       0.35       0.03

    INVERSE MATRIX

         0.66787   -0.02526    0.00000   -0.07294
        -0.01056    0.88073    0.00000   -0.20688
        -0.00148    0.00750    0.39494   -0.33508
        -0.06337   -0.17012    0.00000    2.92953

    SOLUTIONS:
         0.06304    0.07976    0.07536    0.07906

    DETERMINANT VALUE:      0.14944E+01
```

```
LINEAR EQUATIONS SOLUTION ROUTINE

INPUT THE NUMBER OF EQUATIONS IN THE SET
4
INPUT THE   4 CONSTANTS
75. 50. 41. 86.
INPUT THE   4 COEFFICIENTS FOR EQUATION   1
121.0 9.35 1.38 20.2
INPUT THE   4 COEFFICIENTS FOR EQUATION   2
22.4 4.61 74.9 0.0
INPUT THE   4 COEFFICIENTS FOR EQUATION   3
27.1 20.2 1.30 32.8
INPUT THE   4 COEFFICIENTS FOR EQUATION   4
23.0 100.0 6.57 43.8


THE SET OF EQUATIONS:

     121.00        9.35        1.38       20.20      75.00
      22.40        4.61       74.90        0.00      50.00
      27.10       20.20        1.30       32.80      41.00
      23.00      100.00        6.57       43.80      86.00

INVERSE MATRIX

    0.00966     0.00042    -0.00010    -0.00650
    0.00194     0.01383    -0.00091    -0.01967
   -0.00301    -0.00098     0.01344     0.00316
   -0.00906    -0.00882     0.00011     0.04785

SOLUTIONS:
    0.48854     0.48361     0.49169     0.52904

DETERMINANT VALUE:     0.18494E+08
```

REFERENCES

M. L. Boas, *Mathematical Methods in the Physical Sciences*, 2nd ed., Wiley, New York, 1983.

F. E. Hohn, *Elementary Matrix Algebra*, 2nd ed., Macmillan, New York, 1964.

8

RANDOM NUMBERS AND MONTE CARLO SIMULATION

8.1 PSEUDORANDOM NUMBER GENERATION

Random numbers are widely employed in a variety of iterative computer calculations including Monte Carlo studies, experimental design, statistical studies, and many others. Here we will discuss the generation of uniformly distributed (rectangularly distributed) pseudorandom numbers on the unit interval. After generation in this manner, they can be changed to other probability distributions, have their means or variances altered, or have other changes made. Several methods exist for generating pseudorandom sequences, but the most widely used is the multiplicative congruential method, which employs a recurrence relation of the form

$$r_i = (kr_{i-1} + C) \bmod N$$

where the seed number r_0 must lie between 0 and N and the parameters k and C are integers between 0 and $N - 1$. Ordinarily, N is chosen to be very large, typically the largest integer that can be stored in the computer (e.g., $2^{31} - 1 = 2147483647$). With this choice of N, all random numbers generated by the recursive relation will be on the interval between 0 and this value. All multiplicative congruential random number generators cycle, and each integer between 1 and N will appear exactly once per cycle. This makes it desirable for N to be very large. The numbers can be normalized to the unit interval by dividing by N. A specific implementation of this recursive relationship in a FORTRAN subroutine is shown (subroutine RAND).

In the subroutine RAND, the value of N is $2^{31} - 1$, the value for k is $7^5 = 16,807$, and the value for C is 0. The correctness of implementation of RAND on any machine can be checked by verifying that if the seed number 1 is used, then the 1000th value returned is 522,329,230.

The numbers generated by algorithms using the multiplicative congruential method are called *pseudorandom* sequences; they are clearly not true random numbers because they were generated in a deterministic way by a computer program. However, once generated, they are indistinguishable from true random numbers by statistical tests. The two properties that a series of pseudorandom numbers must have to be considered random are (1) uniform distribution and (2) uncorrelated sequence. These can both be tested for using standard statistical methods such as the chi-squared test (e.g., Fluendy 1970). Property 1 refers to the fact that, on the unit interval, there should be just as many members of the random sequence between 0.50 an 0.59 as between 0.10 and 0.19, that is, a rectangular distribution. Property 2 refers to the fact that there should be no discernable relationship between pairs of values in the sequence.

Most random number generators supply rectangularly distributed sequences of values on the unit interval. For many applications, however, random numbers drawn from some other distribution are required. In such cases, the initially supplied values must be transformed. There are several methods available, some general and some very specific. Complete descriptions of available methods can be found in advanced references (e.g., Fluendy 1970, Zelen and Severo 1964).

One particular transformation that is used very often is that from the rectangular distribution to the Gaussian distribution. This transformation can be accomplished simply by taking advantage of the central limit theorem. This theorem states: If a population of variables has a finite variance σ^2 and mean $\bar{x}$, then as the sample size n increases, the distribution of the sample mean $\bar{x}$ approaches the normal distribution with mean $\bar{x}$ and variance σ^2/n. The important property of this theorem is that it states nothing about the form of the population distribution function. Thus, one can take uniform random numbers and use the central limit theorem to generate normal or Gaussian distributed random numbers. In equation form, this reduces to

$$x_k = 2 \sum_{i=1}^{N} e_i - N$$

Values of N in the range 10–12 suffice. Here, a set of uniformly distributed random values e_i plugged into this equation will generate a set of Gaussian distributed values x_k. This method is not as efficient as other methods, but it is easily understood and programmed.*

---

* Function RAND is reprinted with permission from *Transactions of Mathematical Software,* **5,** 132 (1979). Coyright 1979, Association for Computing Machinery.

```
      FUNCTION RAND (IX)
C.... PORTABLE FORTRAN RANDOM NUMBER GENERATOR
C....     FROM A.C.M. TRANS. MATH. SOFTWARE, 5, #2, 132 (1979)
C....       BY LINUS SCHRAGE
C....     USES THE RECURSION   IX = IX * A (MOD P)
C....
C.... INITIALIZE WITH SEED O < SEED < 2**31-1
C....
C.... USE EITHER RAND:  O < RAND < 1
C....          OR IX:  O < IX < 2**31 -1
C....
C.... CHECKING VALUES: IF IX(O)=1, THEN IX(1000)=522329230
C....
C.... IX IN CALLING LIST MUST BE INTEGER*4 IN CALLING PROGRAM
C....
      INTEGER A, P, IX, B15, B16, XHI, XALO, LEFTLO, FHI, K
C.... 7**5, 2**15, 2**16, 2**31-1
      DATA A/16807/, B15/32768/, B16/65536/, P/2147483647/
C....
C.... GET 15 HIGH ORDER BITS OF IX
      XHI = IX/B16
C.... GET 16 LOW BITS OF IX AND FORM LO PRODUCT
      XALO = (IX-XHI*B16)*A
C.... GET 15 HIGH ORDER BITS OF LO PRODUCT
      LEFTLO = XALO/B16
C.... FORM THE 31 HIGHEST BITS OF FULL PRODUCT
      FHI = XHI*A + LEFTLO
C.... GET OVERFLOW PAST 31ST BIT OF FULL PRODUCT
      K = FHI/B15
C.... ASSEMBLE ALL THE PARTS AND PRESUBTRACT P
      IX = (((XALO-LEFTLO*B16) - P) + (FHI-K*B15)*B16) + K
C.... ADD P BACK IN IF NECESSARY
      IF (IX.LT.O) IX = IX + P
C.... MULTIPLY BY 1/(2**31-1)
      RAND = FLOAT (IX) * 4.656612875E-10
      RETURN
      END
```

8.2 MONTE CARLO SIMULATION

The availability to chemists of computer hardware and software of great power has opened up new areas of research where matter or events can be simulated with mathematically based models. The advent of this type of investigation has altered in a fundamental way the approach that many chemists take to studying problems. It opens up the possibility of building quite complex and realistic theoretical models that can be tested against experimental observations. In simulation, we start with simple elements whose behavior is known (such as how molecules move in a potential field), and then we put them together into a larger system (such as simulation of liquids). The objective is to study the complicated system. This section will present a selection of some examples of the types of simulations that have been done in order to show the importance of this area of computer applications in chemistry.

Molecular Dynamics: Modeling of Liquid Water

One extremely important line of research in the area of simulation is the use of molecular dynamics for the simulation of liquid water. In 1971, Rahman and Stillinger published a study that has been cited as the parent of the field. In their study, classical mechanics was invoked. Each water molecule was considered to be a rigid sphere (neon atom) with four partial charges, two negative and two positive, placed at the vertices of a regular tetrahedron so as to represent the known bond electric moments of water molecules. The two partial positive charges simulate partially shielded protons, and the two partial negative charges simulate unshared electron pairs. They constructed a potential that operated on each pair of molecules in the model. The potential consisted of two parts. The first part was the simple Lennard-Jones potential, which would describe the interaction of two neon atoms. This term contained the usual attractive 6th-power term and the standard repulsive 12th-power term:

$$Br_{ij}^{-12} - Ar_{ij}^{-6} \tag{1}$$

The coefficients A and B are parameters of the potential field, and r_{ij} is the distance between water molecules i and j. The second part of the potential field was the simple Coulomb electrostatic potential between the electric charges present:

$$\frac{Cq_iq_j}{r_{ij}} \tag{2}$$

Here C is a parameter of the potential field, and q_i and q_j are the partial charges on the ith and jth water molecules. The potential generated by interactions of the partial charges of one molecule with those of another molecule simulate the hydrogen-bonding tendency of water molecules.

Bulk water was represented by a cubic aggregate of water molecules with six on a side, that is, $6 \times 6 \times 6 = 216$ molecules. The position and orientation of each water molecule was specified by three Cartesian coordinates and three Eulerian angles. In addition, the time derivatives of each of the coordinates and angles were saved to specify the linear and angular momentum of each molecule. Therefore, the total number of parameters being saved was $216 \times 12 = 2592$. At any time during the simulation run, the net force on each of the 216 molecules due to the 215 others was calculated. Then each molecule was allowed to move under the influence of its net force field for a short increment of time ($\Delta t = 4.355 \times 10^{-16}$ sec). The calculations needed to simulate one time increment took about 40 sec, a factor of 10^{17} compared to the motions of real water molecules. Randomized initial states were chosen, and the ensemble of water molecules reached a reasonable temperature (34.3 °C). Then the system was followed in detail for 5000 time increments. A number of water properties were calculated, and stereoscopic views were output for visual comparison. Rahman and Stillinger (1971) reported a number of observations that they made from studying the stereoscopic views: the liquid appeared highly disordered, many OH bonds are not included in hydrogen bonding, few aggregates appear, and so on.

The overall conclusion reached was that classical mechanics applied to a model with only a few hundred molecules was already sufficiently complex to characterize with some degree of accuracy an aggregate as complex as water in the liquid state.

Reaction Dynamics: A Classical Trajectory Calculation

Simulating the motions of atoms and molecules as they undergo either reactive or unreactive collisions is a prominent aspect of modern computational theoretical chemistry. Such simulations are an important complement to experimental studies, which together constitute an active research field known as molecular dynamics (Levine and Bernstein 1974). One of the most powerful simulation procedures is the Monte Carlo classical trajectory method (Bunker 1974). Given the potential energy surface that describes the interactions between the atoms and molecules, one can obtain a complete dynamical microscopic picture of the collision by obtaining the numerical solution of the classical equations of motion. This solution is called a trajectory, and it gives the coordinates, momenta, and energies of all reactant, intermediate, and product species as a function of time during the collision event.

There are four principal steps in a Monte Carlo classical trajectory study. They are:

1. Choosing the potential energy surface for the collision event under study.
2. Selecting initial conditions for the collision partners with a Monte Carlo method so that distributions for the initial conditions represent those for the experimental study with which the trajectory results will be compared.
3. Numerically integrating the classical equations of motion.
4. Transforming the final coordinates and momenta of the trajectories to such properties as angular momenta bond lengths, and bond energies for final analysis of the trajectory results.

The potential energy surface is usually represented by an analytic function that has a set of parameters. Values are chosen for the parameters by fitting experimental data such as vibrational frequencies, bond energies, and activation energies and by fitting theoretical ab initio electronic structure calculations. The ab initio calculations provide detailed information about the shape of the potential energy surface far from equilibrium geometries. An example of an analytic potential energy surface is that for the reaction

$$H + C_2H_4 \longrightarrow CH_5$$

(Hase et al. 1978).

The trajectory obtained from one trajectory is not sufficient for comparison with experiment. For comparison with experimental results, initial conditions are chosen by a Monte Carlo method to represent the experimental conditions under investigation. The Monte Carlo sampling procedures are as diverse as the different types of experimental conditions. For example, the reactants may be in either specific vibrational–rotational states or have a Boltzmann distribution of vibrational and rotational states specified by a temperature T. Normally, the reactants are randomly oriented; however, for some situations it is possible to align the reactants. It is clear that a different Monte Carlo sampling procedure is required for each of these cases.

Standard numerical algorithms are available for integrating the classical equations of motion. Some of the more popular are the Runge–Kutta, Adams–Moulton, and Gear algorithms. In integrating the classical equations of motion, care must be taken to use a sufficiently small time step so that accurate results are obtained. For some collision events, you may want to use an integration algorithm that allows the use of a variable time step. Final trajectory properties such as angular momentum and bond lengths are obtained from the coordinates and moments with standard equations.

To illustrate the Monte Carlo classical trajectory method, details are given of a classical trajectory study by Date, Hase, and Gilbert (1984) of the energy transfer process that occurs when a highly vibrational excited methane molecule collides with an argon atom. The study is performed using a general Monte Carlo classical trajectory computer program named MERCURY, which is available from the Quantum Chemistry Program Exchange as Program No. 453. In this simulation, methane molecules are impinged upon by argon atoms. The methane molecule contains 100 kcal/mole of vibrational–rotational energy, which constitutes a large amount of internal vibrational excitation. The initial relative translational energy of the system is 5 kcal/mole. Thus, the total energy of the system is 105 kcal/mole.

The total energy can be partitioned as

$$E = T + V \tag{3}$$

where E is the total energy, T is the kinetic energy, and V is the potential energy. The potential energy can be written as

$$V = V_{\text{inter}} + V_{\text{intra}} \tag{4}$$

where V_{inter} is the argon–methane ($Ar-CH_4$) intermolecular potential and V_{intra} is the methane intramolecular potential. There are two constants of the motion of the system: the total energy E_t and the total angular momentum J. The total angular momentum is the vector sum $J = j + l$, where j is the methane rotational angular momentum and l is the $Ar-CH_4$ orbital angular momentum.

The intramolecular potential of the methane molecule is represented by four Morse functions (one for each of the four carbon–hydrogen bonds) and five harmonic H–C–H bends. The Morse function has the form

$$V(r) = D\{1 - \exp[-\beta(r - r_0)]\}^2 \tag{5}$$

where the parameters are as follows: r_0 is the internuclear distance for which the energy is minimum, D is the well depth or the dissociation energy, and β is a spectroscopic parameter related to the C–H stretching force constant. The harmonic bends are represented by the functional form

$$V(\theta) = f_\theta(\theta - \theta_0)^2 \tag{6}$$

where θ is the angle of an H–C–H bend, θ_0 is the angle of least strain, and f_θ is the bending force constant.

The Ar–CH$_4$ intermolecular potential is represented as a sum of five Lennard-Jones 6–12 potentials (one for the Ar–C interaction and four for the Ar–H interactions) of the functional form

$$V(r) = 4\epsilon\left[\left(\frac{\sigma}{r}\right)^{12} - \left(\frac{\sigma}{r}\right)^6\right] \tag{7}$$

where ϵ is the depth of the potential well, r is the internuclear distance, and σ is the value of r at which the potential is zero. The Lennard-Jones potential combines an attractive potential with an r^{-6} dependence with a very steeply rising repulsive potential with an r^{-12} dependence.

A Monte Carlo method is used to choose the initial conditions for the trajectories. The vibrations and rotations of methane are assigned random phases with a total energy of 100 kcal/mole. Methane is randomly oriented with respect to the argon by assignment of randomly chosen Euler angles. The collision has a fixed initial impact parameter b of 1 Å, as shown in Figure 1, and an initial relative translational energy of 5 kcal/mole. The impact parameter is the distance of closest approach of the centers of mass of the argon and methane in the absence of any interaction.

Once the initial conditions are chosen, the classical equations of motion are integrated. The numerical integration begins with a fourth-order Runge–Kutta method and, after six cycles, switches to a faster sixth-order Adams–Moulton predictor–corrector algorithm. As the simulation proceeds, the energy E_{vr}, l, and j are calculated after each iteration cycle. The quantity E_{vr} is the vibrational–rotational energy of the methane,

$$E_{vr} = T_{intra} + V_{intra} \tag{8}$$

(T_{intra} is in the center-of-mass frame), l is the Ar–CH$_4$ orbital angular momentum, and j is the methane rotational angular momentum. The orbital angular momentum is given by

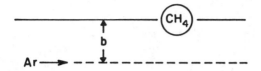

Figure 1. Geometry of collision of argon with methane.

$$l = r_{rel} \times p_{rel} \tag{9}$$

where r_{rel} is the vector of position for the centers of mass of the argon and methane and p_{rel} is the momentum vector for the Ar–CH$_4$ relative motion. Then

$$j = \sum_{i=1}^{s} r_i \times p_i \tag{10}$$

where the r_i and p_i are vectors for the five atoms of methane with respect to the methane center of mass.

The values of the methane vibrational–rotational internal energy, methane rotational angular momentum, and Ar–CH$_4$ orbital angular momentum are plotted as a function of time for a large number of trajectories. The time dependence of these dynamical variables gives one an intimate picture of the collision event. A typical plot is shown in Figure 2. The argon atom and the methane molecule interact strongly for approximately 0.1 psec. The solid line is E_{vr}, and it shows that the changes in the methane molecule's energy during the collision is much larger than the difference between the initial and final values. Strong couplings between the intramolecular motion of the CH$_4$ and the Ar + CH$_4$ intermolecular motion are evident from the modulations in the plots of E_{vr}, l, and j versus time. The plot shown, and many others for different initial conditions, show no evidence for long-lived collision complexes. Each trajectory is found to be characterized by only one inner turning point in the Ar + CH$_4$ relative motion. From studies like this one, the efficiency of energy transfer from the highly excited molecule to the less excited species is found to depend on the structure and intramolecular properties of both species.

The above is only one example of the many different chemical processes that can be simulated by Monte Carlo classical trajectory calculations. Classical trajectories have also provided an in-depth microscopic understanding of bimolecular and unimolecular reactions, intramolecular vibrational energy redistribution, gas–solid surface interactions, and many other types of phenomena. The Monte Carlo classical trajectory method is expected to become even more important for studying molecular reaction dynamics as computer technology advances.

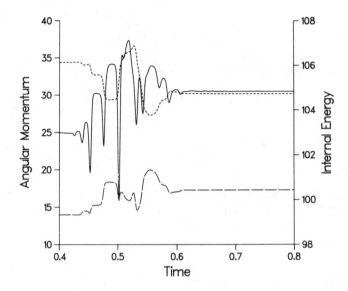

Figure 2. Time-dependent values of CH_4 internal energy E_{vr} (solid line), CH_4 rotational angular momentum (long dashes), and $Ar–CH_4$ orbital angular momentum (short dashes). [Reprinted with permission from *Journal of Physical Chemistry*, **88**, 5135 (1984). Copyright 1984, American Chemical Society.]

Classical Dynamics: Scattering of Ions off Surfaces

An effective way to probe surfaces is to bombard the surface with ions and observe the scattered ions and ejected ions, neutrals, and clusters. When the ejected ions are detected, these experiments are called SIMS (for secondary ion mass spectrometry). SIMS experiments can be simulated using classical mechanics in order to advance understanding of the experimental observations of SIMS in the following way (Garrison and Winograd 1982).

To perform such a study, a model system must be set up. It typically consists of a representation of a segment of metal surface by an array of atoms that is four or five layers deep and on the order of 100 atoms per layer. Different crystal structures or faces of single crystals can be simulated as desired. The interactions between the individual atoms are represented by a suitable classical potential field. The surface of the system is bombarded with an ion with a few kilo-electron-volts of energy. Then, for each impact, one uses classical mechanics to compute the positions and momenta of all the relevant particles as a function of time after the impact of the primary ion. Using classical mechanics means integrating the laws of motion.

Newton's equation of motion for the n particles constituting the simulation is

$$m_i \frac{d^2 \bar{r}_i}{dt^2} = \bar{F}_i(\bar{r}_1, \bar{r}_2, \ldots, \bar{r}_n) \tag{11}$$

where m_i is the mass, $\bar{r}_i$ is the position, and $\bar{F}_i$ is the force on the ith atom. Note that the force depends on the positions of all the atoms and that these positions are time dependent. In principle, one can solve this set of $3n$ coupled second-order differential equations for the position of each atom as a function of time. There are $3n$ equations because there is one equation for the x, the y, and the z direction for each atom. In practice, it is easier to solve an equivalent set of coupled first-order differential equations known as Hamilton's equations of motion. These are given by

$$m_i \frac{d\bar{v}_i}{dt} = \bar{F}_i(\bar{r}_1, \bar{r}_2, \ldots, \bar{r}_n) \tag{12}$$

$$\frac{d\bar{r}_i}{dt} = \bar{v}_i \tag{13}$$

where $\bar{v}_i$ is the velocity of the ith particle. This set of $6n$ equations of motion are mathematically equivalent to the set of $3n$ coupled second-order differential equations. There are two advantages to working with Hamilton's rather than Newton's equations of motion. First, the numerical algorithms for solving coupled first-order differential equations are much better developed. Second, one of the quantities needed as output of the computation is the velocities of the particles in order to compute their kinetic energies and directions of motion, and the velocities are automatically given by Hamilton's equations.

The major problem with any scattering calculation is that the forces acting between the particles must be approximated. For the purposes of the SIMS calculations being disussed here, the forces are assumed to be pairwise additive. That is, each atom interacts with each of its neighbors as if all the remaining atoms were not present. Generally, the force is written in terms of an interaction potential V, where

$$\bar{F}_i = \bar{\nabla}_i \sum_j V(r_{ij}) \tag{14}$$

where r_{ij} is the magnitude of the distance between the ith and the jth particles and $\bar{\nabla}_i$ is the gradient with respect to the ith particle. These pair potentials, $V(r_{ij})$, are picked with the best chemical knowledge available.

In solving Hamilton's equations of motion, one must specify the initial velocities and positions of all the particles. It is generally assumed that the atoms in the solid are initially at their equilibrium configurations and have no initial kinetic energy, that is, they have initial velocities of zero. In the experimental configuration, the primary ion beam is collimated and monoenergetic with the energy E_{ion}. Therefore, its initial velocity v_{ion} is given by

$$v_{ion} = \left(\frac{2E_{ion}}{m_{ion}} \right)^{1/2}$$ (15)

and the components are given by

$$v_z = v_{ion} \cos (\theta)$$ (16)

$$v_x = v_{ion} \sin (\theta) \cos (\phi)$$ (17)

$$v_y = v_{ion} \sin (\theta) \sin (\phi)$$ (18)

where z is the direction perpendicular to the surface, and the x and y directions are in the plane of the surface; θ is the angle with respect to the surface normal, and ϕ is the angle in the plane. The initial z position of the ion is started at a long distance (in particular, 10 Å) above the surface. The x and y positions are randomly chosen so as to mimic the experiment in which ions collide with the surface at random locations. Hamilton's equations are solved, and the calculation is terminated when no more atoms have sufficient energy to be ejected from the surface. Then the final positions and velocities of the atoms are analyzed.

A particle can be ejected from the surface when it has sufficient energy to break away from the attraction of the surface. A surface attraction potential V_{sa} is added to the kinetic energy of a particle to test for ejection,

$$\tfrac{1}{2}mv_z^2 + V_{sa} > 0$$ (19)

If the particle cannot escape the surface, it will migrate along the surface. The ejection yield is determined by averaging the number of particles that eject in each individual ion impact over all the impacts. The value obtained for ejection yield is extremely sensitive to the choice of potential parameters, so the reliabilty of the calculated value is not very good.

For each of the particles that ejects, the kinetic energy is given by

$$\tfrac{1}{2}m(v_x^2 + v_y^2 + v_z^2)$$ (20)

The kinetic energies are averaged and studied.

Several calculations are performed in which ions strike various points on an undamaged surface. The final experimental observables are obtained by averaging over many ion impact points on the surface. Fifty to 3000 impacts might be used. The area on the surface of the system that must be bombarded is determined by the symmetry of the surface being examined. The size of the system studied is chosen so that edge effects are negligible. Because the amount of computer power and time is sometimes finite, care must be taken in choosing which experiments to model. For example, incident particles with lower kinetic energies do not cause as much crystal damage, and thus a smaller crystal is needed for the calculation. The result of the simulation is the final positions and momenta of the particles above the surface. From this information, the mean number of particles ejected per incident primary ion, the kinetic energy, and the angular distributions and cluster formation probabilities can be determined. These quantitites can be compared rather directly to experimental values. In addition to the experimental quantities, the classical dynamics model allows one to examine the collision process in detail. Thus, an understanding at the atomic level can be obtained.

This model can be extended to the study of chemisorbed species on the surface. Such species can be placed on the surface in whatever locations or coverages might be desired, and then the simulation is run in this mode.

At least two important approximations are present in this approach: which potential field to use and how to represent the ionization process. The forces felt by any atom within the system at any time are dependent on all the other atoms. Which of the many potential fields to use depends on the purposes of the study and the preferences of the scientist. A widely used approach is to use pairwise additive potentials that include a repulsion term that dominates at small interatomic distances and a Morse-type potential that dominates at larger interatomic distances. The mean value of particles ejected is quite dependent on the choice of parameters in the potential field. On the other hand, calculated quantities such as energy and angular distributions are rather insensitive to the choice of parameters and can be determined with fairly high accuracy.

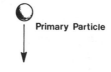

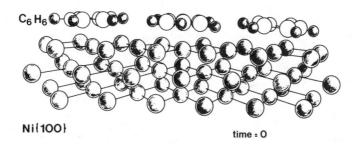

Figure 3. Arrangement of Ni(001) surface before Ar^+ impact. [Reprinted with permission from *Secondary Ion Mass Spectrometry, SIMS IV*, Springer-Verlag, New York, 1984. Copyright 1984, Springer-Verlag.]

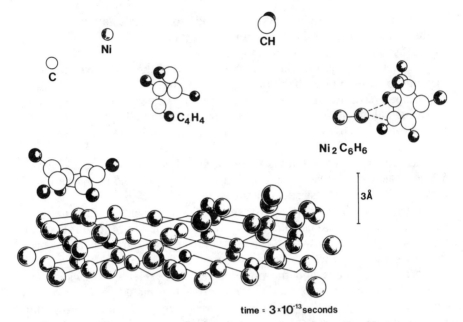

Figure 4. Arrangement of Ni(001) surface and ejected particles 3×10^{-13} sec after Ar^+ impact. [Reprinted with permission from *Secondary Ion Mass Spectrometry*, *SIMS IV*, Springer-Verlag, New York, 1984. Copyright 1984, Springer-Verlag.]

An important process omitted in this and all classical models is the ionization that often occurs during the ejection process. The ionization affects most strongly the mass spectrum. The energy and angular distributions of ions are slightly different from those of the neutral species. The basic features in each are similar, and the classical model can still aid in interpreting the angular distributions of the ions. Several authors have reported methods for including ionization processes in these types of scattering simulations.

A specific example of this type of simulation is given in papers by Garrison (1982, 1984) involving a Ni(001) surfaces covered with benzene molecules being bombarded with energetic argon ions. The predicted mass spectrum resulting from this bombardment was dominated by C_6H_6, in agreement with experiments. The orientation of the benzene molecules on the surface had a profound effect on the results observed. Figures 3 and 4 show before and after snapshots of this simulation. The after picture shows ejected particles and a disarrayed nickel surface at 3×10^{-13} sec after the Ar^+ impact.

Simulation of Chromatographic Processes

When a chemical compound passes through a chromatographic column, a complex series of events occurs. The behavior of the system is determined by the interactions between the compound of interest being carried through the column by the mobile phase and the stationary phase. The sample distributes itself between the two phases dynamically as it passes through the column. A compound that prefers to be adsorbed on the stationary phase to a larger extent will be retained longer than another compound with a lesser such preference. The resulting differential migration of compounds through the column is the basis of chromatographic separations. The fundamental interactions between the compounds being separated and the stationary phase are absorption–desorption processes. The individual absorption and desorption events cannot be observed, however. Only highly averaged quantities such as retention times of eluting compounds and peak shapes are experimentally accessible. These observable quantities are functions of many experimental variables, including mobile phase flow rate, particle size of the solid support, pressure gradient across the column, the uniformity of packing, temperature, and others. The observable quantities are related to many fundamental thermodynamic parameters, including activity coefficients, distribution coefficients, free energies, heats of solution, entropy changes, and others. A mathematical model of the chromatographic processes can be used to probe the connections between the fundamental chemical interactions and the observables. Such models are typically too complex for closed-form solution, so they are used as the starting point for computer simulations.

One approach to simulating chromatographic processes, discrete event simulation, has been studied (Phillips 1981, Wright 1981). This method proceeds by setting up a series of mechanisms for the basic chemical processes involved, each with associated probabilities of occurrence. Random numbers are used to decide which events occur at any given time. Large numbers of events occur, and the simulation collects the results as statistical averages.

In one such study (Phillips 1981), gas–solid chromatography was simulated. The molecules flow in a carrier gas stream over a solid stationary phase. Individual molecules were put through the chromatographic column one at a time. This precludes taking interactions between molecules into account, but it is a reasonable approximation for the dilute concentrations characteristic of the mobile gas phase. Peak shapes were obtained for comparison to experimentally observed chromatographic peaks.

In another study (Wright 1981), the high-performance liquid chromatography (HPLC) column was simulated. Diffusion of the molecules of interest in the mobile phase was neglected as unimportant. The model was focused on describing the surface composition of the stationary phase and the interactions between it and the molecules of interest. The adsorption–desorption events were modeled by a two-step retention mechanism. It was assumed that there was no competition between molecules for adsorption sites, that is, infinite dilution. A molecule in the mobile phase absorbed on an interaction site with some probability. Once there, it could either return directly to the mobile phase or it could jump to a second, nearby adsorption site. Eventually, it returned to the mobile phase The model also provided for injecting molecules into the column, moving them through the column while in the mobile phase, and detecting them upon elution.

The adjustable parameters that were accessible in the model were as follows:

1. Probability of a molecule encountering an interaction site on the stationary phase.
2. Probability of a molecule moving from the first interaction site to the second.
3. Probability of an adsorbed molecule returning to the mobile phase.
4. Sample size, number of interaction sites of each type on the stationary phase.

A series of HPLC experimental studies was done for comparison. Benzene and aniline were put through three stationary phases with differing compositions. Spherisorb was the most polar surface because it had the most silanol groups, Lichrosorb had an intermediate number of silanol groups, and micro-Bondapak C-18 had the least polar surface.

The discrete event simulation was used to compare the calculated versus observed peaks of benzene and aniline as a function of stationary phase polarity. The authors found that their two-site model could produce simulated peaks that could be compared to the experimental ones for analysis. Thus, they showed that HPLC retention could be explained satisfactorily by the two-step retention model. The simulation was an integral part of the overall project, allowing the direct comparison of calculation and observation.

Quadrupole Mass Analyzer Simulation

The quadrupole mass filter (QMF) is a very widely used type of mass spectrometer, and it is incorporated into many commercial instruments. A QMF can scan quickly (1 amu/msec), and this has made it attractive for incorporation into gas chromatograph–mass spectrometer instruments. However, its characteristics are not fully understood even though they have been extensively studied (e.g., Campana 1980). Computer software simulation of the QMF allows investigation of characteristics, such as individual ion trajectories, that are not experimentally observable (Campana and Jurs 1980).

An ideal QMF consists of four long hyperbolic cylinders in a square array with the inside radius of the array equal to the smallest radius of curvature of the hyperbolas. In practice, this geometry is approximated by four parallel cylindrical rods mounted at the corners of a square at a distance of

$2r_0$ from the opposite rod. Opposite rods are electrically connected. If the radius of the rods is made $1.1486r_0$, then the ideal electric field that would be provided by the hyperbolic rods is well approximated by the circular rods. A DC voltage U and an RF voltage $V \cos \omega t$ are impressed between opposite pairs of rods. Voltage V is the peak RF voltage, ω is the angular frequency $2\pi f$, and t is time.

The field potential in the quadrupole field is given by

$$\Phi(x, y, z, t) = \frac{(U + V \cos \omega t)(x^2 - y^2)}{r_0^2} \tag{21}$$

where x, y, and z are the Cartesian coordinates within the field. Since there is no applied potential in the z direction, there is no z term in the field potential. This field potential results in the following electric fields in the x and y directions:

$$E_x = - \frac{2(U + V \cos \omega t)x}{r_0^2} \tag{22}$$

$$E_y = + \frac{2(U + V \cos \omega t)y}{r_0^2} \tag{23}$$

For a certain set of quadrupole operating conditions, a particular ion (characterized by its mass, energy, entrance location, etc.) may have a stable trajectory and survive transit through the QMF. The ions are injected into the field of the QMF parallel to the z axis with an energy determined by the accelerating potential they have fallen through. The ions proceed with a constant velocity in the z direction, and they undergo complicated oscillatory motions in the x and y directions. For a particular set of operating conditions, only ions of a particular m/z will move through the mass analyzer to be collected at the detector. All other ions will collide with the rods and be lost.

The parameters that characterize the analyzer are U, V, $f = \omega/2\pi$, and r_0. Mass scanning is accomplished by varying U and V, keeping their ratio constant, while keeping f constant.

From $F = ma$, the equations of motion of an ion of mass m in the electric fields are as follows:

$$m \frac{d^2x}{dt^2} - eE_x = 0 \tag{24}$$

$$m \frac{d^2y}{dt^2} + eE_y = 0 \tag{25}$$

$$m \frac{d^2z}{dt^2} = 0 \tag{26}$$

The motion of an ion in either direction is independent in an ideal field. These equations are generally used with the following variable substitutions:

$$a = \frac{4eU}{m\omega^2 r_0^2} \tag{27}$$

$$q = \frac{2eV}{m\omega^2 r_0^2} \tag{28}$$

$$\xi = \tfrac{1}{2}\omega t \tag{29}$$

Substitutions of these dimensionless quantities and rearrangement of the equations yields a final equation of the following form for the x-direction motion:

$$\frac{d^2 x}{d\xi^2} = (a_x + 2q_x \cos 2\xi)x = 0 \tag{30}$$

An equation with y replacing x expresses the y-direction motion. These equations are special types of linear second-order differential equations known as *Mathieu* equations. Mathieu equations have been studied intensively and are well understood mathematically (McLachlan 1947).

For certain sets of values for a and q, there are stable solutions to Mathieu equations that correspond to stable ion trajectories through the length of the QMF. Detailed consideration of the a–q space and its region of stability are beyond the scope of this discussion but can be found in the literature (e.g., Campana 1980). Briefly, there is a region in a–q space, bounded by two polynomials, within which the a, q values lead to stable trajectories. The ion trajectories through the QMF are helical; they can be most easily viewed by decomposing the motion into the two orthogonal components in the x–z and the y–z planes. The region of stability is independent of the initial velocity of the ion and the relative phase of the RF field at the time of entry. In practice, a values between 0 and 0.24 and q values between 0 and 0.9 can be used. The exact values of a and q and their ratio determine, to a large extent, the properties of the QMF. The resolution of the QMF ($R = m/\Delta m$) is determined by the value of the ratio a/q.

Trajectories can be found by numerical integration of the differential equations of motion.

Another approach, implemented in the program QMAS (see end of chapter), is to construct a computer model of the QMF. This is a simulation approach in which the electric field is expressed as a function of location and time. The electric field interacts with a charged particle to impart an acceleration that causes a displacement of the ion. The path of an ion can be mapped as it passes through the QMF by repetitive iterations of the above cycle. The results can be viewed on a graphics display terminal.

A program nearly identical to QMAS was used to generate some example ion trajectories reported in a paper by Campana (1980). The QMF parameters used in these simulations were as follows: inscribed radius r_0, 0.277 cm; field length l, 15.24 cm; mass resolution R, 100; RF frequency f, 2.5 MHz; a parameter, 0.2344; q parameter, 0.7037; RF peak voltage V, 690.38 V; DC voltage U, 114.99 V. The ions injected into the field were of mass 100 amu. The x and y entrance coordinates were set to 0.01 cm, that is, the ion entered just off-center to the main axis of the QMF. The x and y velocity components for the ions were 47.053 m/sec. The z velocity component was 3105.496 m/sec. The initial kinetic energy of the ions was 5.0 eV. For these conditions, the transit through the QMR was 4.907×10^{-5} sec. The RF field had a phase angle of zero at the time the ions entered the QMF. During the time the ion was in transit through the field, 122.69 RF cycles occurred. The 100-amu ion was considered to have been formed at a point source 0.660 cm in front of the entrance to the QMF with an initial total energy of 5 eV.

Several different sets of $a-q$ values were chosen, and a trajectory was computed for each. Figures 5–7 show the trajectories resulting from the

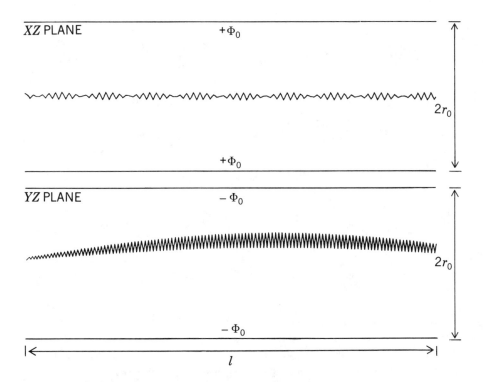

Figure 5. Ion trajectory of the $m/z = 100$ ion calculated for a and q values near the boundary of the $a-q$ stability diagram. [Reprinted with permission from *International Journal of Mass Spectrometry and Ion Physics*, **33**, 101 (1980). Copyright 1980, Elsevier Science Publishers.]

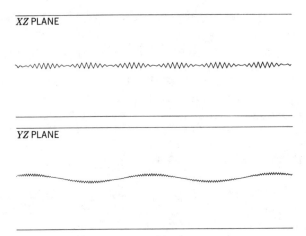

Figure 6. Ion trajectory of the $m/z = 100$ ion calculated for a and q values well within the $a-q$ stability diagram. [Reprinted with permission from *International Journal of Mass Spectrometry and Ion Physics*, **33**, 101 (1980). Copyright 1980, Elsevier Science Publishers.]

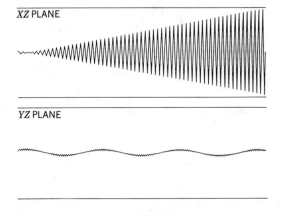

Figure 7. Ion trajectory of the $m/z = 100$ ion calculated for a and q values near the boundary of the $a-q$ stability diagram. [Reprinted with permission from *International Journal of Mass Spectrometry and Ion Physics*, **33**, 101 (1980). Copyright 1980, Elsevier Science Publishers.]

following sets of q, a values: (0.7002, 0.2333), (0.7037, 0.2344), (0.7072, 0.2356). Different sets of a and q result in trajectories that are relatively stable, as in Figure 6, or relatively unstable, as in Figures 5 and 7. Figure 8 shows the trajectory of an ion of mass 99 amu when injected into the QMF tuned to pass only 100-amu ions. Figure 9 shows the trajectory of an ion of mass 101 amu when injected into the same QMF. Since the QMF was tuned to pass ions of mass 100 amu and has a resolution of 100 (unit

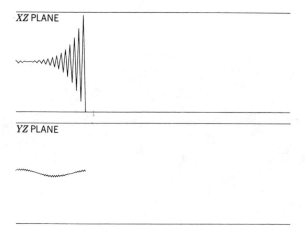

Figure 8. Ion trajectory of the $m/z = 99$ ion calculated at quadrupole operating conditions set to pass only the $m/z = 100$ ion. [Reprinted with permission from *International Journal of Mass Spectrometry and Ion Physics*, **33**, 101 (1980). Copyright 1980, Elsevier Science Publishers.]

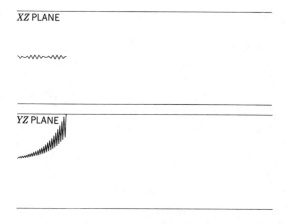

Figure 9. Ion trajectory of the $m/z = 101$ ion calculated at quadrupole operating conditions set to pass only the $m/z = 100$ ion. [Reprinted with permission from *International Journal of Mass Spectrometry and Ion Physics*, **33**, 101 (1980). Copyright 1980, Elsevier Science Publishers.]

resolution at mass 100), the ions with masses of 99 and 101 amu cannot pass all the way through the QMF.

In addition to plotting the individual trajectories of ions passing through the QMF, there are many additional simulation experiments that can be performed. An example is the study of the average properties of ions as they pass through the QMF. The parameters that characterize the interaction between the ion and the electric field within the QMF can be averaged to approximate the behavior of ions in a real laboratory experiment. Such a set

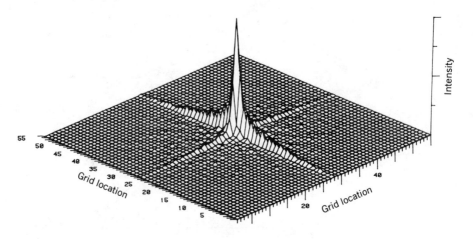

Figure 10. Three-dimensional histogram showing the ion exit distribution from the quadrupole mass analyzer. [Reprinted with permission from *International Journal of Mass Spectrometry and Ion Physics*, **33**, 119 (1980). Copyright 1980, Elsevier Science Publishers.]

of experiments was reported by Campana and Jurs (1980). The full details of the work can be obtained from the original paper. Briefly, the experiment done was as follows. Ions were injected into the QMF in locations spread about the entrance aperture, with varying angles of entrance with respect to the z axis of the QMF, with varying a, q values, with varying RF phase angles, but with fixed energy (5.0 eV). Each study was done with unit resolution. A total of 2600 ions were put through the QMF per simulation. The computation of the trajectory of any ion was terminated if the ion strayed too far from the central axis of the QMF. The location of exit of each ion from the QMF was stored. The fraction of ions surviving transit through the QMF was found to fall off for higher mass ions (at unit resolution). A three-dimensional histogram showing the exit distribution is given in Figure 10. The ions were focused toward the center of the QMF and more along the y axis than the x axis. These results were in accord with experimental observations of QMF behavior.

```
        PROGRAM QMAS
C
C       QUADRUPOLE MASS ANALYZER SIMULATOR
C
        DIMENSION XD(1000),YD(1000)
        COMMON   H,Z,E,ENG
        COMMON /IOUNIT/ NINP,NOUT
        COMMON /IOCTRL/ IOUT
        DATA NOE/'N'/,PI/3.14159/,IEND/'E'/,IYES/'Y'/
        DATA NBAUD/4800/
        NINP=1
        NOUT=1
 1001 CONTINUE
        WRITE (NOUT,209)
```

```
  209 FORMAT (' THIS IS PROGRAM QMAS',/)
      WRITE (NOUT,219)
  219 FORMAT (' ENTER THE DESIRED RESOLUTION')
      READ (NINP,*) RES
      WRITE (NOUT,229)
  229 FORMAT (' ENTER MINIMUM AND MAXIMUM MASSES IN A.M.U. ')
      READ (NINP,*) MMIN,MMAX
      WRITE (NOUT,239)
  239 FORMAT (' CHOOSE OUTPUT DESIRED: ',/,10X,'1 FOR PRINT ONLY',/,
     X     10X,'2 FOR TRAJECTORY DRAWINGS AS WELL')
      READ (NINP,*) IOUT
      IF (IOUT.GT.2) IOUT=2
      IF (IOUT.LT.1) IOUT=1
      IF (IOUT.EQ.2) CALL INITT (NBAUD)
C     CON IS THE NUMBER OF ITERATIONS PER RF CYCLE
      CON=100.0
      RZ=0.277
      F=1.8
      RATAQ=2.*(.16784-.12588/RES)
      CALL INTR (QL,QR,RATAQ)
      I=0
 1003 CALL TRAJEK(QL,QR,MMIN,MMAX,RATAQ,RES,RZ,N,A,Q,XZ,YZ,VX,VY,VZ,F,
     1TT,U,V,G,I,MFREK,CON)
      XZS=XZ
      YZS=YZ
      VXS=VX
      VYS=VY
      JNUMB=1
      NDEN=1
      IF (NDEN.EQ.0) NDEN=1
      KK=1
      DO 1010 JJ=1,JNUMB
      XZ=XZS
      YZ=YZS
      VX=VXS
      VY=VYS
      CALL QUAD(XD,YD,VX,VY,VZ,TT,F,U,V,RZ,G,N,XZ,YZ,R,XMAX,YMAX,
     XIPLT,JJ,A,Q,RATAQ,MFREK,NDEN)
      IF (IOUT.EQ.2) CALL PLOT (XD,YD,RZ,MFREK,KK,IPLT)
      IF (IOUT.EQ.2) CALL TSEND
      IF (IOUT.EQ.2) READ (NINP,249) IANS
  249 FORMAT (A1)
      KK=2
      IF (IOUT.NE.2) GOTO 1010
      WRITE (NOUT,259)
  259 FORMAT (' WANT AN END-VIEW PLOT?')
      READ (NINP,249) IANS
      IF (IANS.EQ.NOE) GO TO 1011
      IF (IANS.NE.IYES) GO TO 1010
      IF (IOUT.EQ.2) CALL PLTEND(XD,YD,RZ,IPLT)
      IF (IOUT.EQ.2) CALL TSEND
      IF (IOUT.EQ.2) CALL HOME
      IF (IOUT.EQ.2) CALL TSEND
 1010 CONTINUE
      WRITE (NOUT,269)
  269 FORMAT (' CONTINUE ?')
      READ (NINP,249) IANS
 1011 CALL ERASE
      CALL TSEND
      WRITE (NOUT,279)
  279 FORMAT (' ANOTHER RUN ?',///,' 1 START OVER',/,' 2 RESET QUAD',/,
     1' 3 INJECT ANOTHER ION',/,' 4 STOP')
      READ (NINP,*) III
      IF (III.EQ.1) GO TO 1001
      I=1
      IF (III.EQ.2) GO TO 1003
      I=3
      IF (III.EQ.3) GO TO 1003
      STOP
      END
```

```
C---------------------------------------------------------------
      SUBROUTINE TRAJEK(QL,QR,MMIN,MMAX,RATAQ,RES,RZ,N,A,Q,XZ,YZ,VX,VY,
     1 VZ,F,TT,U,V,G,I,MFREK,CON)
C.... INPUT ION ENTRANCE CONDITIONS; PERHAPS SET UP QUAD TO FOCUS IT
      COMMON /IOUNIT/ NINP,NOUT
      COMMON /IOCTRL/ IOUT
      DATA PI/3.14159/,EL/15.24/,PSD/0.660/
      WRITE (NOUT,109)
  109 FORMAT (' ENTER ION MASS IN A.M.U. ')
      READ (NINP,*) G
      WRITE (NOUT,119)
  119 FORMAT (' ENTER ION ENERGY IN ELECTRON VOLTS')
      READ (NINP,*) E
      WRITE (NOUT,129)
  129 FORMAT (' ENTER OFF AXIS X AND Y ENTRY LOCATIONS')
    6 READ (NINP,*) XZ,YZ
      IF (XZ.LT.RZ.AND.YZ.LT.RZ) GOTO 7
      WRITE (NOUT,139)
  139 FORMAT (' TOO FAR FROM CENTER ... TRY AGAIN')
      GOTO 6
    7 VT=SQRT((2.0*E*1.60219E-12)/(G*1.66042E-24))
      THETA=ATAN(YZ/XZ)
      R=SQRT(XZ**2+YZ**2)
      PHI=ATAN(R/PSD)
      PHI=180.0*PHI/PI
      THETA=180.0*THETA/PI
      VV=VT*SIN(PHI*PI/180.0)
      VX=VV*COS(THETA*PI/180.0)
      VY=VV*SIN(THETA*PI/180.0)
      VZ=VT*COS(PHI*PI/180.0)
      IF (I .EQ. 3) GO TO 1
      MDIFF=MMAX-MMIN
      QDIFF=QR-QL
      QINC=QDIFF/MDIFF
      Q=QL+QINC*(G-MMIN)
      A=Q*RATAQ
      U=1.66042*A*PI**2*G*F**2*RZ**2/1.60219
      V=2.*U/RATAQ
      WRITE (NOUT,149)
  149 FORMAT (/,' ENTRY CONDITIONS')
      WRITE (NOUT,159) A,Q,RATAQ,V,U,RES
  159 FORMAT (' A = ',F12.5,/,' Q = ',F12.5,/,' A/Q = ',F10.5,/,
     X  ' V =',F10.2,/,' U = ',F9.2,/,' RESOLUTION = ',F5.0)
    1 TT=EL/VZ
      N=TT*F*1.0E+06*CON
      CNOS=TT*F*1.0E+06
      MFREK=CNOS*2.0
      WRITE (NOUT,169)  CNOS,VX,VY,VZ,THETA,PHI,TT
  169 FORMAT (' NO. OF RF CYCLES = ',F7.2,/,' ENTRY VELOCITIES = ',
     X   3F7.0,/,' ENTRY ANGLES (THETA,PHI) = ',2F7.1,/,
     X   ' EXPECTED TIME IN RF FIELD = ',E12.4,' SEC.')
      RETURN
      END
C---------------------------------------------------------------
      SUBROUTINE QUAD(XD,YD,VX,VY,VZ,TT,F,U,V,RZ,G,N,XZ,YZ,R,XMAX,YMAX,
     X XIPLT,JJ,A,Q,RATAQ,MFREK,NDEN)
C.... PUT AN ION THROUGH THE QUADRUPOLE MASS FILTER
      DIMENSION XD(1000), YD(1000)
      COMMON /IOUNIT/ NINP,NOUT
      COMMON /IOCTRL/ IOUT
      DATA PI/3.14159/
      YMAX=0.0
      XMAX=0.0
      IF (JJ .NE. 1) MFREK=MFREK-1
      XM=1.0/(G*1.66042E-24)
      RZS=RZ**2
      FRAC=FLOAT(JJ-1)/FLOAT(NDEN)
      M=0
      KK=1
      XD(KK)=XZ
```

```
      YD(KK)=YZ
      NO=N/MFREK
      DT=TT/N
      TZ=FRAC/(F*1.0E+06)
      TWOPI=2.0*PI*F*1.0E+06
      ITR=0
      DO 20 I=1,N
      T=DT*(I-1)
      FAC=(U+V*COS(TWOPI*(T-TZ)))/RZS
      EX=-FAC*XZ
      EY=FAC*YZ
      FX=1.60219E-12*EX
      FY=1.60219E-12*EY
      AX=FX*XM
      AY=FY*XM
      DX=VX*DT+(AX*DT**2)/2.0
      DY=VY*DT+(AY*DT**2)/2.0
      XZ=DX+XZ
      YZ=DY+YZ
      M=M+1
      IF (M-NO) 11,8,11
    8 KK=KK+1
      XD(KK)=XZ
      YD(KK)=YZ
      M=0
   11 XTEM= ABS(XZ)
      YTEM= ABS(YZ)
      IF (XTEM.GT.XMAX) XMAX=XTEM
      IF (YTEM.GT.YMAX) YMAX=YTEM
      VX=VX+AX*DT
      VY=VY+AY*DT
      IF (XTEM.GE.RZ.OR.YTEM.GE.RZ) GO TO 24
   20 CONTINUE
      ITR=1
   24 KK=KK+1
      XD(KK)=XZ
      YD(KK)=YZ
      IPLT=KK
      R=SQRT(XZ**2+YZ**2)
      MFREK=MFREK+1
      IF (JJ.NE.1) RETURN
      THETA=ATAN(VY/VX)
      VV=SQRT(VY**2+VX**2)
      PHI=ATAN(VV/VZ)
      PHI=180.0*PHI/PI
      THETA=180.0*THETA/PI
      WRITE (NOUT,109)
  109 FORMAT (/,' EXIT CONDITIONS')
      IF (ITR.EQ.0) WRITE (NOUT,119)
  119 FORMAT (' ION NOT TRANSMITTED')
      IF (ITR.EQ.1) WRITE (NOUT,129) VX,VY,VZ,THETA,PHI
  129 FORMAT (' EXIT VELOCITIES = ',3F10.0,/,' EXIT ANGLES ',
     X  '(THETA,PHI) = ',2F10.3)
      IF (ITR.EQ.1) WRITE (NOUT,139) IPLT,XZ,YZ,R,XMAX,YMAX
  139 FORMAT (' NO. OF POINTS PER TRAJECTORY = ',I5,/,
     X  ' EXIT LOCATION (X,Y,R) = ',3F10.3,/,
     X  ' MAX. DISTANCES FROM AXIS (X,Y) = ',2F10.3)
      WRITE (NOUT,149)
  149 FORMAT (' CONTINUE ?')
      READ (NINP,159) IANS
  159 FORMAT (A1)
      RETURN
      END
C----------------------------------------------------------------
      SUBROUTINE INTR (QL,QR,RATAQ)
C     BISECTION METHOD OF  ROOT FINDING
C     FIND INTERSECTIONS OF SCAN LINE WITH EDGES OF STABILITY DIAGRAM
      DATA  ACC/1.0E-05/
      SCAN(Q,RATAQ)=Q*RATAQ
```

```
      A(Q)=(.5*Q**2)-(0.05469*Q**4)+(0.01259*Q**6)-(0.00364*Q**8)
      B(Q)=1.0-Q-(0.12500*Q**2)+(0.01563*Q**3)-(0.00065*Q**4)-
     1(.00034*Q**5)+(.00025*Q**6)
      FUNC1(Q,RATAQ)=A(Q)-SCAN(Q,RATAQ)
      FUNC2(Q,RATAQ)=B(Q)-SCAN(Q,RATAQ)
      DO 100 II=1,2
      L=II-1
      A1=0.0
      A2=0.706
      IF (L.EQ.1) A1=1.0
      I=0
      IF(L.EQ.0) GO TO 40
      B1=FUNC2(A1,RATAQ)
      B2=FUNC2(A2,RATAQ)
      GO TO 50
40    B1=FUNC1(A1,RATAQ)
      B2=FUNC1(A2,RATAQ)
50    IF (B1) 51,51,86
51    IF (B2) 86,52,52
52    X=(A1+A2)/2.
      I=I+1
      IF (L.EQ.0) GO TO 60
      Y=FUNC2(X,RATAQ)
      GO TO 80
60    Y=FUNC1(X,RATAQ)
80    IF (ABS(X-A2)-ACC) 86,86,83
83    IF (Y) 84,86,85
84    A1=X
      GO TO 52
85    A2=X
      GO TO 52
86    Y=RATAQ*X
      IF (L.EQ.0) QL=X
      IF (L.EQ.1) QR=X
100   CONTINUE
      RETURN
      END


C------------------------------------------------------------------
      SUBROUTINE PLOT(XD,YD,RZ,N,JJ,I)
C.... PLOT TRAJECTORIES OF IONS IN XY AND XZ PLANES
      DIMENSION  XD(1000),YD(1000)
      COMMON /IOUNIT/ NINP,NOUT
      COMMON /IOCTRL/ IOUT
      INTEGER*2 NN10,NN20,NN350,NN370,NN410,NN760,NN1010,NN1020
      INTEGER*2 NN2,ICH1,ICH2
      DATA NN10/10/,NN20/20/,NN350/350/,NN370/370/,NN760/760/
      DATA NN1010/1010/,NN1020/1020/,NN410/410/,NN2/2/
      DATA ICH1/'XZ'/,ICH2/'YZ'/
      IF (JJ.NE.1) GO TO 5
      CALL ERASE
      CALL MOVABS(NN10,NN760)
      CALL TSEND
      CALL AOUTST (NN2,ICH1)
      CALL TSEND
      TRZ=2.*RZ
      XN=FLOAT(N)
      CALL MOVABS (NN10,NN410)
      CALL DRWABS(NN1010,NN410)
      CALL MOVABS(NN10,NN760)
      CALL DRWABS(NN1020,NN760)
      CALL VWINDO(0.0,XN,-RZ,TRZ)
5     CALL SWINDO(NN10,NN1010,NN410,NN350)
      CALL MOVEA(0.0,XD(1))
      DO 10 J=2,I
      W=FLOAT(J)
      CALL DRAWA(W,XD(J))
10    CONTINUE
```

```
      IF (JJ.NE.1) GO TO 15
      CALL MOVABS(NN10,NN370)
      CALL TSEND
      CALL AOUTST (NN2,ICH2)
      CALL TSEND
      CALL MOVABS(NN10,NN370)
      CALL DRWABS(NN1020,NN370)
      CALL MOVABS(NN10,NN20)
      CALL DRWABS(NN1020,NN20)
   15 CALL SWINDO(NN10,NN1010,NN20,NN350)
      CALL MOVEA(0.0,YD(1))
      DO 20 J=2,I
      W=FLOAT(J)
      CALL DRAWA(W,YD(J))
   20 CONTINUE
      RETURN
      END
C-------------------------------------------------------------
      SUBROUTINE PLTEND (XD,YD,RZ,I)
C.... PLOT END-ON VIEW OF ION PATH THROUGH QUADRUPOLE
      DIMENSION XD(1000),YD(1000)
      INTEGER*2 NNO,NN780
      DATA NNO/0/,NN780/780/
      CALL ERASE
      CALL SWINDO(NNO,NN780 ,NNO,NN780)
      RZN=-RZ
      RZT=2.0*RZ
      CALL VWINDO (RZN,RZT,RZN,RZT)
      CALL MOVEA (RZN,0.0)
      CALL DRAWA (RZ ,0.0)
      CALL MOVEA (0.0,RZN)
      CALL DRAWA (0.0,RZ )
      CALL MOVEA (XD(1),YD(1))
      DO 10 K=3,I,2
      CALL DRAWA (XD(K),YD(K))
   10 CONTINUE
      RETURN
      END
      THIS IS PROGRAM QMAS
      ENTER THE DESIRED RESOLUTION
      100
      ENTER MINIMUM AND MAXIMUM MASSES
      99 101
      ENTER THE VALUE FOR THE SCALING CONSTANT
      500
      ENTER ION MASS IN A.M.U.
      100
      ENTER ION ENERGY IN ELECTRON VOLTS
      5
      ENTER X AND Y ENTRY LOCATIONS
      0.05 0.10


      A,Q,RATAQ  0.23446    0.70373    0.33316
      V,U      357.88        59.62
      RES      100.


      NO. OF CYCLES      89.57
      VX,VY,VZ     23202.      46405.      306270.
      THETA,PHI   63.435        9.615
      TT     0.4976E-04
      ENTER NUMBER OF IONS AND FRACTION
      1 1
      VX,VY,VZ    -16784.     351178.      306270.
      PHI,THETA   48.940      -87.264
      IPLT        8
      XZ,YZ,R    -0.006       0.277        0.277
```

```
XMAX, YMAX     0. 050     0. 277
CONTINUE?
N
    plot comes out here
CONTINUE ?
N
    plot comes out here


1 START OVER
2 RESET QUAD
3 INJECT ANOTHER ION
4 STOP
4
```

LIST OF VARIABLES USED IN PROGRAM QMAS

Quadrupole Mass Filter Operating Conditions

MMIN, MMAX	Minimum and maximum masses of ions to be used
RES	Resolution
RATAQ	Ratio of parameters a–q; scan line slope
QL, QR	Values of q where scan line intersects the left and right edges of the stability diagram
RZ	Inscribed circle radius in the quadrupole, cm
F	Frequency of the RF field, MHz
U	DC voltage
V	RF peak voltage
EL	Length of quadrupole rods, cm
PSD	Distance from point source to quadrupole entrance aperture, cm

Ion Characteristics

G	Mass of injected ion, amu
E	Energy of injected ion, eV
XZ, YZ	x and y entrance locations, cm
VT	Velocity of injected ion
PHI, THETA	Entrance angles of injected ion
VX, VY, VZ	Velocities in the x, y, and z directions
T	Transit time of ion through quadrupole
N	Number of evaluations of force field to be made
CNOS	Number of cycles of RF field seen by ion during transit through field
EX, EY	Electric fields in x and y directions
FX, FY	Forces in x and y directions
AX, AY	Accelerations in x and y directions
DX, DY	Displacements in x and y directions
XD, YD	Storage of positions of ion at time increments

Other Parameters

MFREK	Number of points to be plotted

REFERENCES

General

D. L. Bunker, "Simple Kinetic Models from Arrhenius to the Computer," *Accts. Chem. Res.*, **7**, 195–201 (1974).

M. Fluendy, "Monte Carlo Studies," in *Markov Chains and Monte Carlo Calculations in Polymer Science*, G. G. Lowry (Ed.), Marcel Dekker, New York, 1970, Chapter 3.

P. G. Lykos, *Computer Modelling of Matter*, American Chemical Society, Washington, DC, 1978.

L. Schrage, "A More Portable Fortran Random Number Generator," *A. C. M. Trans. Math. Software*, **5**, 132 (1979).

M. Zelen and N. C. Severo, "Methods of Generating Random Numbers and Their Applications," in *Handbook of Mathematical Functions with Formulas, Graphs, and Mathematical Tables*, M. Abramowitz and I. A. Stegun (Eds.), National Bureau of Standards Applied Mathematics Series, No. 55, U.S. Government Printing Office, Washington, DE, 1964, Section 26.8, pp. 949–953.

Water Simulation

A. Rahman and F. H. Stillinger, "Molecular Dynamics Study of Liquid Water," *J. Chem. Phys.*, **55**, 3336 (1971).

Classical Trajectory

N. Date, W. L. Hase, and R. G. Gilbert, "Collisional Deactivation of Highly Vibrationally Excited Molecules. Dynamics of the Collision Event," *J. Chem. Phys.*, **88**, 5135–5138 (1984).

W. L. Hase, "MERCURY: A General Monte Carlo Classical Trajectory Program," *Q.C.P.E.*, **3**, 453 (1983).

W. L. Hase, G. Mrowka, R. J. Brudzynski, C. S. Sloane, "An Analytical Function Describing the $H + C_2H_4 = C_2H_5$ Potential Energy Surface," *J. Chem. Phys.*, **69**, 3548–3562 (1978).

R. D. Levine and R. B. Bernstein, *Molecular Reaction Dynamics*, Oxford University Press, New York, 1974.

Surface Scattering

B. J. Garrison, "Organic Molecule Ejection from Surface due to Heavy Particle Bombardment," *J. Amer. Chem. Soc.*, **104**, 6211 (1982).

B. J. Garrison, "Mechanisms of Organic Molecule Ejection in SIMS and FABMS Experiments," in *Secondary Ion Mass Spectroscopy, SIMS IV*, A. Benninghoven, J. Okano, R. Shimizu, and H. W. Werner (Eds.), Springer-Verlag, Berlin, 1984.

B. J. Garrison and N. Winograd, "Ion Beam Spectroscopy of Solids and Surfaces," *Science*, **216**, 805 (1982).

Chromatographic Simulation

J. B. Phillips, N. A. Wright, and M. F. Burke, "Probabilistic Approach to Digital Simulation of Chromatographic Processes," *Sep. Sci. Technol.* **16**, 861 (1981).

N. A. Wright, "*Computer Assisted Investigations of Chromatographic Processes*," Masters Thesis, University of Arizona, 1981.

Quadrupole Mass Filter Simulation

J. E. Campana, "Elementary Theory of the Quadrupole Mass Filter," *Int. J. Mass. Spectr. Ion Phys.*, **33**, 101–117 (1980).

J. E. Campana and P. C. Jurs, "Computer Simulation of the Quadrupole Mass Filter," *Int. J. Mass Spectr. Ion Phys.*, **33**, 119–137 (1980).

P. H. Dawson, Ed., *Quadrupole Mass Spectrometry and Its Applications*, Elsevier, Amsterdam, 1976.

N. W. McLachlan, *Theory and Application of Mathieu Functions*, Oxford University Press, New York, 1947.

9

SIMPLEX OPTIMIZATION

9.1 THE SIMPLEX METHOD

Many activities of chemists involve observing the output or response of a system, whether it is an instrument or a mathematical model or a reaction, as a function of a number of experimental variables. For example, the common activity of tuning an instrument means adjusting the instrumental settings in order to seek the best response. The best response may be the best sensitivity or the best selectivity or some combination of both.

In the development of new analytical chemical methods and in the improvement or extension of existent ones, it is common to investigate the effects of experimental variables on the results. Variables such as pH, temperature, reagent concentration, and so on, must be varied to find those that have an effect on the observed results. Then the values of these important variables are optimized to improve the results of the method. Improvement can mean increased selectivity, increased sensitivity, decreased interferences, faster operation, greater precision, or some other desired outcome.

The use of chemical instrumentation nearly always involves tuning the instrument for the best response for the experimental measurement being made. This is a classical optimization problem, especially since the instrumental settings are ordinarily interrelated. Consider a specific example, atomic absorption spectroscopy with a flame atomizer. Experimental variables that are directly related to the observed signal level and precision of measurements include the hollow cathode lamp current, widths of the various slits in the monochromator, flame conditions, fuel and oxidant flow rates, burner position, spectral line used, integration time used for detection, and many others as well.

In organic synthesis, the conditions are varied in order to seek the best yield of the desired product. All of these activities, and many more, involve optimization. A good, workable definition of optimization has appeared as follows: "the collective process of finding the best set of conditions required to achieve the best result from a given situation" (Beveridge and Schechter 1970).

The first obstacle to optimization of real systems is the multidimensional nature of the problem; that is, there are usually many variables to optimize simultaneously. If the variables that are being investigated are independent of one another, it is a straightforward task to vary them one at a time while observing the response. Independent variables can be optimized individually. However, in real situations, such complete independence is rarely found. The variables to be considered are almost always related to one another so that one must study them collectively. Multivariate methods must be used.

There are several classes of optimization methods, several of which we will discuss here. Four of them are single-factor variation, grid searches, random approaches, and the simplex method. The single-factor variation method is well suited to the case of complete independence among the variables being investigated. However, the method is often applied for convenience to cases where the variables are interdependent even though it is not really applicable. Grid searches involve the evaluation of the response for several different sets of values of the variables chosen so that they form a grid in the factor space. Once a region of the factor space is located that has good response values, the grid size can be decreased in order to focus in on the best region. Then the process can be repeated with ever smaller grid sizes. The random approach involves trying widely varying values for the variables so that the factor space is sampled far and wide. This method can be wasteful in that a great deal of time can be spent evaluating responses in uninteresting regions of the factor space. The fourth optimization method listed above is the simplex method. Here random variation is replaced with an orderly, statistical design. Evaluations of the response of the system are made, and then this information is used to determine the direction to move for the next evaluation. Then the cycle is repeated, and the overall operation is driven quite efficiently by the use of all the information that is available at any given time.

The Simple Simplex Method

The simplex method is a sequential optimization method that involves repeated observation of the system response, selection of new values for the variables, followed by another observation, and so on. The method can be visualized for optimization problems involving just two variables as a method for tracking on the response surface where one axis corresponds to variable 1, a second axis corresponds to variable 2, and the third axis corresponds to the response. This corresponds to seeking hills or ridges on a topographical map. A geometric construct called a *simplex* is used as the method for tracking about on the response surface.

A simplex is a geometric figure with its number of vertices equal to one more than the number of dimensions of the factor space. A simplex in two dimensions is a triangle; in three dimensions it is a tetrahedron. In four or more dimensions, simplexes cannot be drawn or readily visualized, but the geometry is intact and the method is applicable. So, in general, the number of vertices of a simplex is the number of dimensions in the factor space (equal to the number of variables being investigated) plus 1. Figure 1 shows a simplex in a two-dimensional factor space with its vertices labeled B, N, and W.

A new simplex can be constructed adjacent to an existing one by retaining all but one vertex of the existing one and creating just one new vertex. The simplex in Figure 1 can be converted into an adjacent one by eliminating any one of the three vertices and placing a new vertex anywhere in the factor space. If the new vertex were placed so that it was colinear with the two other vertices, the simplex would have lost one of its dimensions, but this is a special case that we can neglect.

Given that an existent simplex can be changed to another one that is adjacent to the original one, we can make a simplex "walk" about in factor space by a series of replacements. Note that each move requires only one evaluation or observation of the response. To start a simplex investigation of a factor space requires $n + 1$ observations to establish the original simplex. Henceforth, each move requires just one additional observation. The simplex moves about in the factor space quite efficiently, sampling the response values in an orderly way.

To this point, we have not addressed the main question, namely, deciding which direction to move the simplex. We wish to have the simplex move from its current location toward a better region of the factor space. We have

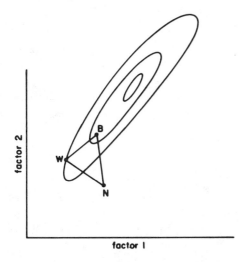

Figure 1. Simplex on response surface.

the current values of the response where the simplex is now located to work with, and this is all we need. The following paragraphs discuss how the simplex moves through the factor space. The discussion follows that of Deming and Morgan (1973).

Rule 1. The simplex is moved after each observation of the response. Once an initial simplex exists, a move can be made after each additional observation.

Rule 2. A move is made into the adjacent simplex obtained by discarding the vertex of the current simplex corresponding to the least desirable response; it is replaced with its mirror image across the face of the simplex, consisting of the remaining vertices. In Figure 2, the original simplex has its vertices labeled B, N, and W. The B vertex corresponds to the best response, N to the next-to-the-worst response, and W to the worst response. Since W is the worst response, it is discarded, and a new vertex is constructed. The new vertex is found along the line segment connecting W with the centroid of the simplex face of the remaining vertices, here C. The new vertex is on the line segment connecting W and C and is placed an equal distance beyond C as W is from C. It is labeled R for reflection vertex (Figure 2).

After the application of rule 2, several outcomes are possible. If the response at R is better than that at B or N, rule 2 can be applied again, and another adjacent simplex will be found. This loop can continue until an optimum is found. However, if the response at R is the worst response in the new simplex, application of rule 2 would put the simplex back where it

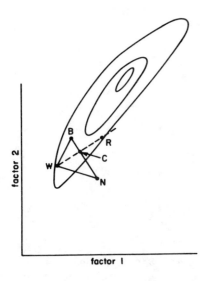

Figure 2. Simplex with labeled vertices, including the reflection vertex.

factor 2

factor 1

just came from, and no improvement would be forthcoming, as it would be trapped in an oscillation between two adjacent simplexes. Accordingly, a way to break out of this oscillation is necessary.

Rule 3. If the new vertex found by reflection has the least desirable response in the new simplex, rule 2 does not apply; but to continue the series, the second lowest response in the new simplex is rejected. This rule allows the simplex to turn as it samples the response surface. However, oscillations can still occur because the simplex can move about a loop in several moves. Thus, another rule is needed to eliminate this possibility.

Rule 4. If a particular vertex has been retained in $k + 1$ simplexes, where k is the dimension of the simplexes, the value of the response at this persistent vertex is noted. This is an attempt to ensure that an anomalous observation has not entered into the optimization.

Rule 5. If a new vertex is outside the boundaries of the allowable factor values, a response observation is not made, but instead a very undesirable response value is assigned to this vertex. This will force the simplex back into the allowable region of factor space so that the optimization can continue.

There are some problems with the simple simplex optimization procedure embodied in the rules stated above. How can one tell when an optimum is reached? What size should the initial simplex be? How can one be sure that the optimum reached is global? None of these questions have definitive answers. One strategy to use in order to home in on the optimum region of the factor surface would be to run an optimization with a certain size simplex and then start over in the vicinity of the optimum with a smaller simplex. This will work, but it is not the best way to proceed. In practice, the simple simplex method is not used because of practical difficulties. Some modified methods that skirt some of these problems have been developed, and we discuss them next.

The Variable-Size Simplex Method

Great improvement in the simplex optimization method can be achieved by allowing the simplex to expand and contract as it probes the response surface. Nelder and Mead (1965) developed the variable-size simplex method, which has been called the modified simplex method (MSM) in the literature. The expansions or contractions are done during the construction of the adjacent simplex from the existent one in the following way. (See Figure 3.) Given a simplex labeled with B, N, and W as before, construct the reflection vertex R. Now, evaluate the response at R, and do one of the following operations depending on the value of the response at R.

If the response at R is better than that of B, the simplex is heading in a

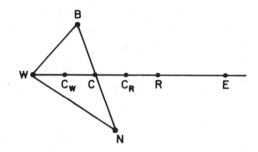

Figure 3. Location of expansion, reflection, and concentration vertices of a simplex.

desirable direction, so try going farther in this same direction. Go to an expansion vertex E. Usually, the distance from C to E is taken as twice that from C to R. If the response at the expansion vertex E is better than R, save it, and the new simplex is BNE. If the response at E is not better than R, keep BNR as the new simplex.

If the response at R is between that of B and N, the reflection vertex R is kept, and the new simplex is BNR.

If the response at R is between that of N and W, a contraction is called for because the simplex is headed in the wrong direction. Depending on how bad the response at R is found to be, the contraction vertex can be placed midway between W and C or midway between R and C. If $N > R > W$, the new vertex is C_R. If $R < W$, the new vertex is C_W.

Using these rules, the simplex can grow and contract as it moves about the factor space searching for the optimum. It is much less likely to become trapped in local optima than is the fixed-size simplex. All that is left is to discuss the termination criteria. How does one decide when to stop the optimization procedure?

A number of convergence criteria are possible. The one to be used should be chosen by the user with the knowledge of the chemical system involved. Some possible convergence criteria are as follows:

1. Based on the values of the response found—The absolute change or relative change in the response must fall below a threshold. This amounts to looking for level parts of the response surface. In addition to the top of peaks, however, there can be other level portions of a surface, for example, inflection points and saddles.
2. Based on the values of the factors—The absolute or relative changes in the values of the factors can be used. This amounts to allowing the simplex size to determine the stopping criterion.
3. Based on small gradients—Numerical evaluations of gradients can be used, but saddle points can fool this type of stopping criterion.

4. Model fitting to several values of the response to try to get a picture of the response surface.

None of these convergence criteria guarantee that the optimization has found a global optimum. None guarantee convergence. In practice, it is usual to follow several criteria simultaneously and to make judgments during the optimization procedure.

9.2 CHEMICAL APPLICATIONS OF SIMPLEX OPTIMIZATION

Ernst (1968) was the first to use the simplex method in analytical chemistry in his optimization of NMR magnetic field homogeneity. Gradient and curvature settings of the instrument were varied in a two-dimensional simplex with good results.

Chubb, Edward, and Wong (1980) applied simplex optimization to increasing the yield of the Bucherer–Burgs reaction. The reaction is a complex one:

$$R_2C{=}O + NH_3 + COS + HCN \longrightarrow R_2C\begin{matrix} N-C=O \\ | \\ C-N \\ \| \\ S \end{matrix} + H_2O$$

At least eight variables bear on the yield obtained—the initial concentrations of ketone, ammonia, HCN, and carbonyl sulfide plus pH, temperature, and time of the reaction and the solvent. The mixed solvent of ethanol–water was used, with the ratio varied as part of the optimization. Chubb et al. (1980) reported the results of several sets of experiments using cyclohexanone or adamantanone as the starting material. They achieved rapid progress in improving the yields of the reaction using the variable-size simplex optimization procedure. In one series of runs, they improved the yield with cyclohexanone from 49 to 88%. They compared several alternative optimization strategies for their applicability toward organic synthesis.

Morgan and Deming (1975) applied simplex optimization to chromatographic methods development. The goal was the separation of five isomeric octanes, 2,3-dimethylhexane, 3-methylheptane, 2,2-dimethylhexane, 2,2,3,3-tetramethylbutane, and 3,3-dimethylhexane. Packed column gas chromagotraphy was used, and the two experimental variables under control of the simplex algorithm were column temperature and carrier gas flow rate. A chromatographic resolution function was used as the response function for the experiments. Dramatic increases in the quality of the chromatograms were seen.

Routh, Swartz, and Denton (1977) used simplex optimization in a flame spectroscopy experiment. Calcium emission at 422.7 nm was observed in a nitrous oxide–acetylene flame. The objective was to maximize the net emission signal due to calcium as a function of the following experimental variables: vertical position of burner, horizontal position of burner, fuel flow rate, oxidant flow rate, monchromator setting, monochromator slit width, and photomultiplier high-voltage setting. Rapid convergence to the optimum settings of the variables was observed with several alternative forms of the simplex method, including a variation proposed by the authors called the super modified simplex.

Leary and co-workers (1982) employed the variable-size sequential simplex method to optimize the instrumental operating conditions for an inductively coupled plasma spectrometer. A multiple-element analysis for Al, Na, Ti, P, and Mn was done with a direct-reading polychromator optically coupled to an argon plasma source. The forward power and the observation height were varied during the simplex optimization. The last simplex found was the best overall for the determination of the five elements, although it did not correspond to the best conditions for any single element.

Berridge (1982) reported the use of the modified sequential simplex algorithm for the unattended optimization of reversed-phase HPLC separations. A chromatographic response function (CRF) was evaluated to represent individual chromatograms with respect to resolution and time of analysis. The CRF contained a number of terms that were derived from the experimentally observed chromatographs. The experiments utilized a microprocessor-controlled liquid chromatograph. Three different sets of experiments were done using two or three variables under investigation. In the first experiment, four 2-substituted pyridines were separated; in the second, a set of three phenolic antioxidants were separated; in the third, a set of four substituted phenols was separated. The studies showed that completely unattended HPLC optimizations were practical.

Harper and co-workers (1983) reported using a simplex search to locate an optimum for an ultrasonic extraction of trace elements from atmospheric particulates collected on glass-fiber high-volume sample filters. A method was found that is quantitative for 13 elements and has been designated as an U.S. Environmental Protection Agency reference method. The four experimental variables that were optimized were the ratio of hydrochloric to nitric acid, ultrasonic bath temperature, ultrasonic duration, and elapsed time of a diffusion step involved in removing the trapped material from the filter. Studies of the accuracy and precision of recovery of the following elements were reported: As, Ba, Cd, Co, Cr, Cu, Fe, Mn, Mo, Ni, Pb, Sr, Ti, V, and Zn. The method is being used to collect data for achiving for all of these elements except Co, Cr, Sr, and Ti.

9.3 NONLINEAR LEAST-SQUARES DATA FITTING BY SIMPLEX

Although the simplex optimization method is well suited to experimental optimization, it can also be used with good results for the fitting of data by nonlinear equations. This is done by setting up the problem so that the response surface that the simplex will investigate is the error function of the data-fitting problem. This is exactly the same error function used in describing curve fitting, namely

$$Q = \sum_{i=1}^{n} [y_i - f(x_i)]^2$$

The parameter whose values are to be found are contained within the function $f(x)$ in the equation. An advantage of this approach to data fitting is that there is no penalty paid if the function is nonlinear in the parameters being fit.

Program SIMPLX

Program SIMPLX, (see end of chapter) implements the simplex method of optimization for the problem of fitting a set of data with a given function. The program is designed to be executed on a time-sharing computer under the immediate supervision of the user. The program begins by asking the user to input the initializing information and the set of data being fit. Then it proceeds to search for the best values of the parameters.

   The function subroutine named ERROR evaluates the error function when called by the main routine. In this implementation of the simplex procedure, the better response is the smaller value since an error is being minimized. The statement in which the variable YCALC is computed is where the program must be changed in order to use a different function from that used in this example. The function being used here as an example is

$$A = A_\infty(1 - e^{-kt})$$

where A_∞ and k are to be found using simplex optimization.

   Program SIMPLX is set up as an accommodation between flexibility and simplicity. Several of the program parameters have been defined within the program, although they could be supplied at execution time by the user if desired. For example, MAXCNT is the number of error function evaluations that are allowed before the program will abort execution on the assumption that no further progress is possible. ERRMIN is given a value of 1.0×10^{-3} inside the program, but this too could be supplied by the user at execution time if desired. The program has several extra WRITE statements in it that will only be executed if the user answers yes to the debug output question.

The stopping criterion implemented in program SIMPLX is the following. When the fractional change in response surface values between the best and worst vertex in the current simplex falls below the value in ERRMIN, the routine terminates.

The set of data being fit in the example execution of the program is taken from a literature paper by Deming and Morgan (1973). The values of the data set are as follows:

t	A	t	A
1.5	0.110	9.0	0.325
1.5	0.109	12.0	0.326
3.0	0.169	12.0	0.330
3.0	0.172	15.0	0.362
4.5	0.210	15.0	0.383
4.5	0.210	18.0	0.381
6.0	0.251	18.0	0.372
6.0	0.255	24.0	0.422
9.0	0.331	24.0	0.411

Program SIMLX was executed twice with this same set of data but with different starting values for the two parameters in the equation. In the first run, the starting values were provided as 0.5 and 1.0. The value of the error function for this set of parameter values was 0.8194. As can be seen from the output of this run, the error value rapidly decreased as the simplex procedure homed in on the best values. A great deal of the output has been deleted in the interest of saving space. The final results obtained were $A_\infty = 0.404$, $k = 0.170$, and an error value of 0.0036, which are all identical with the values reported in the original paper. In the second run, the starting values were provided as 1.0 and 0.5. The starting value for the error function was 7.009 in this case. Again, the error rapidly decreased as the simplex homed in on the best values. The final results were identical with the first run in all respects.

```
      PROGRAM SIMPLX
C....
C....  SIMPLEX MINIMIZATION OF A FUNCTION
C....
      DIMENSION C(10),E(10),P(10,10),R(10),X(10)
      DIMENSION DATA(100,10)
C....  ENTER INITIAL INFORMATION
      MAXCNT=500
      ERRMIN=1.0E-03
      NOUT=1
      NINP=1
      NSIM=1
      WRITE (NOUT,1)
    1 FORMAT (' THIS IS PROGRAM SIMPLX',/)
      WRITE (NOUT,3)
    3 FORMAT (' ENTER NUMBER OF OBSERVATIONS')
```

```
      READ (NINP,*) NOBS
      WRITE (NOUT,4)
    4 FORMAT (' ENTER NUMBER OF VARIABLES (X+Y)')
      READ (NINP,*) NV
      WRITE (NOUT,5)
    5 FORMAT (' ENTER X1,X2,...,XP AND Y')
      DO 6 I=1,NOBS
    6 READ (NINP,*) (DATA(I,J),J=1,NV)
      IF (NOBS.GT.NV) GO TO 11
      WRITE (NOUT,9)
    9 FORMAT (' NO. OF OBS. MUST BE GT NO. OF VARS. ')
      STOP
   11 WRITE (NOUT,12)
   12 FORMAT (' ENTER NUMBER OF PARAMETERS')
      READ (NINP,*) NP
      NP1=NP+1
      WRITE (NOUT,13)
   13 FORMAT (' ENTER INITIAL ESTIMATES OF PARAMETERS')
      READ (NINP,*) (X(I),I=1,NP)
      E(1)=ERROR(X,DATA,NV,NOBS,KOUNT)
      WRITE (NOUT,15) E(1)
   15 FORMAT (' STARTING ERROR FUNCTION VALUE',G12.4)
      WRITE (NOUT,17)
   17 FORMAT (' WANT DEBUG LEVEL OUTPUT? (Y OR N)')
      READ (NINP,18) ANS
   18 FORMAT (A1)
      IDB=0
      IF (ANS.EQ.'Y') IDB=1
C.... INITIALIZE THE SIMPLEX
      KOUNT=0
      DO 22 J=1,NP
   22 P(1,J)=X(J)
      DO 28 I=2,NP1
         DO 26 J=1,NP
   26    P(I,J)=X(J)
         P(I,I-1)=1.1*X(I-1)
         IF (ABS(X(I-1)).LT.1.0E-12) P(I,I-1)=0.0001
   28 CONTINUE
C.... FIND PLOW AND PHIGH/ BEST=PLOS/ WORST=PHIGH
   31 ILO=1
      IHI=1
      DO 34 I=1,NP1
         DO 32 J=1,NP
   32    X(J)=P(I,J)
         E(I)=ERROR(X,DATA,NV,NOBS,KOUNT)
         IF (E(I).LT.E(ILO)) ILO=I
         IF (E(I).GT.E(IHI)) IHI=I
   34 CONTINUE
      WRITE (NOUT,36)
   36 FORMAT (/,' INITIAL SIMPLEX')
      DO 40 K=1,NP1
      WRITE (NOUT,39) K,E(K),(P(K,J),J=1,NP)
   39 FORMAT (3X,' VERTEX',I2,' ERROR AND PARAMETERS: ',5F8.3)
   40 CONTINUE
C.... FIND PNHI THE NEXT HIGHEST   NEXT=PNHI
   41 NHI=ILO
      DO 43 I=1,NP1
         IF (E(I).GE.E(NHI).AND.I.NE.IHI) NHI=I
   43 CONTINUE
C.... COMPUTE THE CENTROID
      DO 46 J=1,NP
         C(J)=-P(IHI,J)
         DO 44 I=1,NP1
            C(J)=C(J)+P(I,J)
   44    CONTINUE
         C(J)=C(J)/NP
   46 CONTINUE
   51 CONTINUE
```

```
C.... PRINT CURRENT BEST VERTEX
      WRITE (NOUT,53) KOUNT,NSIM
   53 FORMAT (' AFTER',I3,' ERROR EVALUATIONS AND',I3,' SIMPLEXES')
      WRITE (NOUT,54) (P(ILO,J),J=1,NP)
   54 FORMAT ('     PARAMETER ESTIMATES: ',5G12.4)
      WRITE (NOUT,55) E(ILO)
   55 FORMAT ('     ERROR FUNCTION: ',G12.4)
C.... STOPPING CRITERION
      IF (KOUNT.GT.MAXCNT) STOP
      IF (ABS(E(ILO)-E(IHI))/E(ILO).LT.ERRMIN) GO TO 56
      GO TO 61
   56 WRITE (NOUT,57)
   57 FORMAT (/,'==> ERROR CRITERION SATISFIED')
      WRITE (NOUT,54) (P(ILO,J),J=1,NP)
      STOP
C.... REFLECTION
   61 DO 62 J=1,NP
         R(J)=1.9985*C(J)-0.9985*P(IHI,J)
   62 CONTINUE
      ER=ERROR(R,DATA,NV,NOBS,KOUNT)
      IF (IDB.GT.0) WRITE (NOUT,65) ER,(R(J),J=1,NP)
   65 FORMAT (' REFLECTION VERTEX',3F10.5)
C.... REFLECT AGAIN IF SUCCESSFUL
      IF (ER.LT.E(ILO)) GO TO 91

      IF (ER.GE.E(IHI)) GO TO 122
C.... REPLACE WORST VERTEX WITH NEW ONE
   79 DO 80 J=1,NP
         P(IHI,J)=R(J)
   80 CONTINUE
      NSIM=NSIM+1
      E(IHI)=ER
      IF (ER.GT.E(NHI)) GO TO 51
      IHI=NHI
      GO TO 41
C.... EXPAND THE SIMPLEX
   91 ILO=IHI
      IHI=NHI
      DO 93 J=1,NP
         X(J)=1.95*R(J)-0.95*C(J)
   93 CONTINUE
      EX=ERROR(X,DATA,NV,NOBS,KOUNT)
      IF (EX.LT.ER) GO TO 104
C.... R BETTER THAN X
      DO 99 J=1,NP
         P(ILO,J)=R(J)
   99 CONTINUE
      NSIM=NSIM+1
      E(ILO)=ER
      GO TO 110
C.... X IS BETTER THAN R
  104 DO 105 J=1,NP
         P(ILO,J)=X(J)
  105 CONTINUE
      IF (IDB.GT.0) WRITE (NOUT,106) EX,(X(J),J=1,NP)
  106 FORMAT (' EXPANSION VERTEX',3F10.5)
      NSIM=NSIM+1
      E(ILO)=EX
  110 CONTINUE
      GO TO 41
C.... CONTRACT THE SIMPLEX
  122 DO 123 J=1,NP
         R(J)=0.5015*C(J)+0.4985*P(IHI,J)
  123 CONTINUE
      ER=ERROR(R,DATA,NV,NOBS,KOUNT)
      IF (IDB.GT.0) WRITE (NOUT,124) ER,(R(J),J=1,NP)
  124 FORMAT (' CONTRACTION VERTEX',3F10.5)
         IF (ER.LT.E(ILO)) GO TO 91
      IF (ER.LT.E(IHI)) GO TO 79
```

```
C.... SCALE
      WRITE (NOUT,135)
  135 FORMAT (' ENTER SCALE (<0 EXPANDS, >0 SHRINKS, 0=STOP): ')
      READ (NINP,*) SCAL
      IF (SCAL.EQ.0.0) GO TO 999
  137 DO 138 I=1,NP1
          DO 138 J=1,NP
              P(I,J)=P(I,J)+SCAL*(P(ILO,J)-P(I,J))
  138 CONTINUE
      GO TO 31
  999 STOP
      END
C------------- --------- - ------------ ----- ----------------
      FUNCTION ERROR (X,DATA,NV,NOBS,KOUNT)
C.... COMPUTES THE ERROR FUNCTION FOR THE DATA SET
C....     SMALLER VALUE IS BETTER
      DIMENSION X(10),DATA(100,10)
      ERROR=0.0
      DO 10 I=1,NOBS
          YOBS=DATA(I,NV)
C.... CHANGE THE NEXT STATEMENT TO CHANGE THE FUNCTION BEING FIT
          YCALC=X(1)*(1.0-EXP(-X(2)*DATA(I,1)))
          RESI=YOBS-YCALC
          ERROR=ERROR+RESI*RESI
   10 CONTINUE
      KOUNT=KOUNT+1
      RETURN
      END

      THIS IS PROGRAM SIMPLX

      ENTER NUMBER OF OBSERVATIONS
      18
      ENTER NUMBER OF VARIABLES (X+Y)
      2
      ENTER X1,X2,...,XP AND Y
      1.5 0.110
      1.5 0.109
      3.0 0.169
      3.0 0.172
      4.5 0.210
      4.5 0.210
      6.0 0.251
      6.0 0.255
      9.0 0.331
      9.0 0.325
      12.0 0.326
      12.0 0.330
      15.0 0.362
      15.0 0.383
      18.0 0.381
      18.0 0.372
      24.0 0.422
      24.0 0.411
      ENTER NUMBER OF PARAMETERS
      2
      ENTER INITIAL ESTIMATES OF PARAMETERS
      0.5 1.0
      STARTING ERROR FUNCTION VALUE  0.8194
      WANT DEBUG LEVEL OUTPUT? (Y OR N)
      N

      INITIAL SIMPLEX
          VERTEX 1 ERROR AND PARAMETERS:   0.819   0.500   1.000
          VERTEX 2 ERROR AND PARAMETERS:   1.204   0.550   1.000
          VERTEX 3 ERROR AND PARAMETERS:   0.848   0.500   1.100
      AFTER  3 ERROR EVALUATIONS AND  1 SIMPLEXES
          PARAMETER ESTIMATES:  0.5000      1.000
          ERROR FUNCTION:  0.8194
```

```
AFTER   5 ERROR EVALUATIONS AND   2 SIMPLEXES
     PARAMETER ESTIMATES:   0.4026        1.147
     ERROR FUNCTION:   0.3372
AFTER   7 ERROR EVALUATIONS AND   3 SIMPLEXES
     PARAMETER ESTIMATES:   0.3565        1.022
     ERROR FUNCTION:   0.1894
AFTER   9 ERROR EVALUATIONS AND   4 SIMPLEXES
     PARAMETER ESTIMATES:   0.2594        1.170
     ERROR FUNCTION:   0.1615
AFTER  10 ERROR EVALUATIONS AND   5 SIMPLEXES
     PARAMETER ESTIMATES:   0.2594        1.170
     ERROR FUNCTION:   0.1615
AFTER  13 ERROR EVALUATIONS AND   6 SIMPLEXES
     PARAMETER ESTIMATES:   0.2608        1.070

     ERROR FUNCTION:   0.1558
AFTER  16 ERROR EVALUATIONS AND   7 SIMPLEXES
     PARAMETER ESTIMATES:   0.3082        1.071
     ERROR FUNCTION:   0.1362
AFTER  18 ERROR EVALUATIONS AND   8 SIMPLEXES
     PARAMETER ESTIMATES:   0.3096        0.9724
     ERROR FUNCTION:   0.1287
AFTER  20 ERROR EVALUATIONS AND   9 SIMPLEXES
     PARAMETER ESTIMATES:   0.3096        0.9724

              MUCH OUTPUT DELETED HERE

AFTER  64 ERROR EVALUATIONS AND  30 SIMPLEXES
     PARAMETER ESTIMATES:   0.4029        0.1724
     ERROR FUNCTION:   0.3620E-02
AFTER  67 ERROR EVALUATIONS AND  31 SIMPLEXES
     PARAMETER ESTIMATES:   0.4064        0.1681
     ERROR FUNCTION:   0.3616E-02
AFTER  70 ERROR EVALUATIONS AND  32 SIMPLEXES
     PARAMETER ESTIMATES:   0.4050        0.1686
     ERROR FUNCTION:   0.3605E-02
AFTER  72 ERROR EVALUATIONS AND  33 SIMPLEXES
     PARAMETER ESTIMATES:   0.4050        0.1686
     ERROR FUNCTION:   0.3605E-02
AFTER  73 ERROR EVALUATIONS AND  34 SIMPLEXES
     PARAMETER ESTIMATES:   0.4050        0.1686
     ERROR FUNCTION:   0.3605E-02
AFTER  76 ERROR EVALUATIONS AND  35 SIMPLEXES
     PARAMETER ESTIMATES:   0.4038        0.1702
     ERROR FUNCTION:   0.3604E-02

==> ERROR CRITERION SATISFIED
     PARAMETER ESTIMATES:   0.4038        0.1702


     THIS IS PROGRAM SIMPLX

     ENTER NUMBER OF OBSERVATIONS
     18
     ENTER NUMBER OF VARIABLES (X+Y)
     2
     ENTER X1, X2, ... , XP AND Y
     1.5 0.110
     1.5 0.109
     3.0 0.169
     3.0 0.172
     4.5 0.210
     4.5 0.210
     6.0 0.251
     6.0 0.255
     9.0 0.331
     9.0 0.325
     12.0 0.326
     12.0 0.330
     15.0 0.362
     15.0 0.383
     18.0 0.381
     18.0 0.372
```

```
24. 0 0. 422
24. 0 0. 411
 ENTER NUMBER OF PARAMETERS
2
 ENTER INITIAL ESTIMATES OF PARAMETERS
1. 0 0. 5
 STARTING ERROR FUNCTION VALUE   7. 009
 WANT DEBUG LEVEL OUTPUT? (Y OR N)
N

 INITIAL SIMPLEX
    VERTEX 1 ERROR AND PARAMETERS:    7. 009   1. 000   0. 500
    VERTEX 2 ERROR AND PARAMETERS:    9. 209   1. 100   0. 500
    VERTEX 3 ERROR AND PARAMETERS:    7. 256   1. 000   0. 550
 AFTER   3 ERROR EVALUATIONS AND  1 SIMPLEXES
    PARAMETER ESTIMATES:   1. 000      0. 5000
    ERROR FUNCTION:   7. 009
 AFTER   5 ERROR EVALUATIONS AND  2 SIMPLEXES
    PARAMETER ESTIMATES:   0. 8053     0. 5737
    ERROR FUNCTION:   3 805
 AFTER   7 ERROR EVALUATIONS AND  3 SIMPLEXES
    PARAMETER ESTIMATES:   0. 7131     0. 5112
    ERROR FUNCTION:   2. 402
 AFTER   9 ERROR EVALUATIONS AND  4 SIMPLEXES
    PARAMETER ESTIMATES:   0. 2903     0. 6251
    ERROR FUNCTION:   0. 9657E-01
 AFTER 10 ERROR EVALUATIONS AND  5 SIMPLEXES
    PARAMETER ESTIMATES:   0. 2903     0. 6251
    ERROR FUNCTION:   0. 9657E-01
 AFTER 12 ERROR EVALUATIONS AND  6 SIMPLEXES
    PARAMETER ESTIMATES:   0. 2903     0. 6251

    ERROR FUNCTION:   0. 9657E-01
 AFTER 14 ERROR EVALUATIONS AND  7 SIMPLEXES
    PARAMETER ESTIMATES:   0. 2903     0. 6251
    ERROR FUNCTION:   0. 9657E-01
 AFTER 16 ERROR EVALUATIONS AND  8 SIMPLEXES
    PARAMETER ESTIMATES:   0. 2903     0. 6251
    ERROR FUNCTION:   0. 9657E-01
 AFTER 19 ERROR EVALUATIONS AND  9 SIMPLEXES
    PARAMETER ESTIMATES:   0. 2940     0. 5901
    ERROR FUNCTION:   0. 9006E-01
 AFTER 22 ERROR EVALUATIONS AND 10 SIMPLEXES
    PARAMETER ESTIMATES:   0. 3264     0. 5905
    ERROR FUNCTION:   0. 8529E-01


            MUCH OUTPUT DELETED HERE

 AFTER 64 ERROR EVALUATIONS AND 31 SIMPLEXES
    PARAMETER ESTIMATES:   0. 4031     0. 1705
    ERROR FUNCTION:   0. 3608E-02
 AFTER 66 ERROR EVALUATIONS AND 32 SIMPLEXES
    PARAMETER ESTIMATES:   0. 4031     0. 1705
    ERROR FUNCTION:   0. 3608E-02
 AFTER 68 ERROR EVALUATIONS AND 33 SIMPLEXES
    PARAMETER ESTIMATES:   0. 4031     0. 1705
    ERROR FUNCTION:   0. 3608E-02
 AFTER 69 ERROR EVALUATIONS AND 34 SIMPLEXES
    PARAMETER ESTIMATES:   0. 4031     0. 1705
    ERROR FUNCTION:   0. 3608E-02
 AFTER 70 ERROR EVALUATIONS AND 35 SIMPLEXES
    PARAMETER ESTIMATES:   0. 4031     0. 1705
    ERROR FUNCTION:   0. 3608E-02
 AFTER 71 ERROR EVALUATIONS AND 36 SIMPLEXES
    PARAMETER ESTIMATES:   0. 4031     0. 1705
    ERROR FUNCTION:   0. 3608E-02
 AFTER 73 ERROR EVALUATIONS AND 37 SIMPLEXES
    PARAMETER ESTIMATES:   0. 4031     0. 1705
    ERROR FUNCTION:   0. 3608E-02
 AFTER 74 ERROR EVALUATIONS AND 38 SIMPLEXES
    PARAMETER ESTIMATES:   0. 4031     0. 1705
```

```
ERROR FUNCTION:   0.3608E-02
AFTER 77 ERROR EVALUATIONS AND 39 SIMPLEXES
  PARAMETER ESTIMATES:   0.4035      0.1710
  ERROR FUNCTION:   0.3606E-02

==> ERROR CRITERION SATISFIED
  PARAMETER ESTIMATES:   0.4035      0.1710
```

REFERENCES

J. C. Berridge, "Unattended Optimization of Reversed-Phase High-Performance Liquid Chromatographic Separations Using the Modified Simplex Algorithm," *J. Chromat.*, **244**, 1 (1982).

G. S. G. Beveridge and R. S. Schechter, *Optimization Theory and Practice*, McGraw-Hill, New York, 1970.

F. L. Chubb, J. T. Edward, and S. C. Wong, "Simplex Optimization of Yields in the Bucherer-Bergs Reaction," *J. Org. Chem.*, **45**, 2315–2320 (1980).

W. K. Dean, K. J. Heald, S. N. Deming, "Simplex Optimization of Reaction Yields," *Science*, **189**, 805–806 (1975).

S. N. Deming and S. L. Morgan, "Simplex Optimization of Variables in Analytical Chemistry," *Anal. Chem.*, **45**, 278A–283A (1973).

S. N. Deming and S. L. Morgan, "Advances in the Application of Optimization Methodology in Chemistry," in *Chemometrics: Theory and Application*, B. R. Kowalski (Ed.), American Chemical Society, Washington, DC, 1977.

R. R. Ernst, "Measurements and Control of Magnetic Field Homogeneity," *Rev. Sci. Inst.*, **39**, 998–1012 (1968).

D. M. Fast, P. H. Culbreth, E. J. Sampson, "Multivariate and Univariate Optimization Studies of Liquid-Chromatographic Separation of Steroid Mixtures," *Clin. Chem.*, **28**, 444–448 (1982).

S. L. Harper, J. F. Walling, D. M. Holland, and L. J. Pranger, "Simplex Optimization of Multielement Ultrasonic Extraction of Atmospheric Particulates," *Anal. Chem.*, **55** , 1553 (1983).

J. J. Leary, A. E. Brookes, A. F. Dorrzapf, Jr., D. W. Gologhtly, "An Objective Function for Optimization Techniques in Simultaneous Multiple-Element Analysis by Inductively Coupled Plasma Spectrometry," *Appl. Spectrom.*, **36**, 37 (1982).

D. E. Long, "Simplex Optimization of the Response from Chemical Systems," *Anal. Chim. Acta.* **46**, 193 (1969).

D. L. Massart, A. Dijkstra, and L. Kaufman, *Evaluation and Optimization of Laboratory Methods and Analytical Procedures*, 2nd ed., Elsevier, Amsterdam, 1980.

S. L. Morgan and S. N. Deming, "Optimization Strategies for the Development of Gas–Liquid Chromatographic Methods," *J. Chrom.*, **112**, 267–285 (1975).

J. A. Nelder and R. Mead, "A Simplex Method for Function Minimization," *Comput. J.*, **7**, 308–313 (1965).

M. W. Routh, P. A. Swartz, and M. B. Denton, "Performance of the Super Modified Simplex," *Anal. Chem.*, **49**, 1422 (1977).

B. G. M. Vandeginste, "Optimization of Analytical Information," *Trends in Analytical Chemistry*, **1**, 210–215 (1982).

PART III

NONNUMERICAL METHODS

Digital computers are universal information processing machines, and as such they can do much more than perform numerical computations. Part III of this book will focus on tasks that can be performed by computer software that are primarily nonnumerical. That is, the tasks, while they must involve the handling of numbers, have as their main focus the performance of some other purpose that is largely nonnumerical.

10

CHEMICAL STRUCTURE INFORMATION HANDLING

10.1 INTRODUCTION

The least common denominator of communication between chemists regarding molecular structures of compounds is the standard two-dimensional structural diagram. Blackboard sketching is an art form in chemistry. This type of sketched structure provides visual communication of concepts that would be extremely cumbersome to describe orally. So much imbedded within chemistry is this "language" that if you draw a hexagon for a chemist with no explanation at all, and ask what it is, you are likely to get the reply, "cyclohexane." Imagine an organic chemistry text barred from using structural diagrams, and the importance of this "language" is even more apparent. These two-dimensional structural diagrams are an abstraction, and they have very little relation to the fundamental nature of the species they represent. They are a shorthand description, not an explanation.

However, standard two-dimensional structural diagrams for chemical compounds are inadequate for many operations of modern chemistry. Examples are abundant, and journals commonly print structures in different ways meant to suggest the three-dimensional aspects of structure. Especially for the input, storage, manipulation, and display of chemical structures within computer-based systems, some more powerful representation must be chosen. It must be compatible with the chemist users' knowledge and background, and it must simultaneously be compatible with the demands of computers. A number of different representations have been developed over the past 20 years. The selection of a representation to be used in a specific set of circumstances depends on many factors, including the following ones: the size of the files to be handled (hundreds or tens of thousands of compounds), the functions to be performed on the files (sort and print lists, prepare indexes, substructure searching), the available hardware (micro-, mini-, or mainframe computer), the available software, the degree of automation desired, and the knowledgeabiity of the personnel using and maintaining the system. The method of structural representation chosen is critical to much of the remainder of the system in a myriad of ways.

Two important characteristics of a structural representation are (1) uniqueness, that is, only one representation can be derived from a compound, and (2) unambiguousness, a representation applies to only one compound. These two characteristics are complementary, since one applies to encoding and the other to decoding of a structural representation. Additionally, a representation must be complete and therefore describe the entire structural diagram. For practical reasons, it is also desirable for a representation to be concise so that it uses as little storage space as possible.

One scheme for classifying structural representations of molecules is given in Figure 1. Fragment codes represent structures by noting the presence of predetermined selected portions of them, such as functional groups, ring systems, and so on. They are ambiguous and are mainly important only as the historical antecedents of the present systems. The unambiguous representations of several classes will be dealt with here in detail.

Topological representations record only the topology or connectivity of the molecule. That is, they record the atom types, bond types, and interconnections among the atoms. Extensions can also record absolute configurations or stereochemistry about asymmetric centers. There are two main classes of topological representations: linear notations and connection tables.

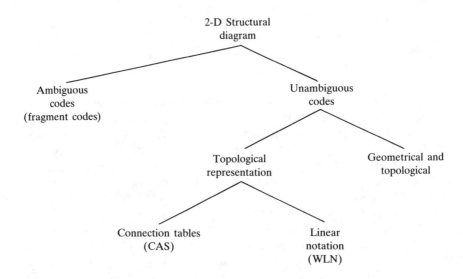

Figure 1. Chemical structure representations. (Reprinted with permission from *Computer Handling of Chemical Structure Information.* Macdonald, London, 1971. Copyright 1971, Macdonald.)

Linear notations condense three-dimensional molecular structures into a unique, unambiguous one-dimensional string of symbols. The notations are compact (an advantage when stored) because most of the bonds are not recorded explicitly, but implicitly. The notation, the sequence of symbols, for a given structure is produced by following a set of rules. Linear notations were originally designed for manual use, but they were readily adaptable for use with punched-card sorting equipment (late 1940s) and computers. While a number of linear notations have been developed (Wiswesser, IUPAC, Hayward), the Wiswesser line notation (WLN) has been the most widely adopted.

10.2   WISWESSER LINE NOTATION

The objective of WLN is to generate a unique code for any chemical structure consisting of a string of letters and numbers that can be written, or typed, on one line. The bonds are usually implied in the string of symbols rather than being cited explicitly.

WLN uses 40 symbols, the 26 capital letters, the 10 numerals, and 4 others: b, &, /, and _. The symbols b or the underscore will be used in this book to denote the space as a WLN symbol. All of these symbols are available on an ordinary typewriter, keypunch, and computer terminal keyboards. These symbols have different purposes, depending on context. First, they can represent individual atoms or functional groups. Second, they can perform syntactical functions. The purpose being served by a particular symbol within a citation is always clear from the context in which it arises.

WLN differentiates among the ways in which heteroatoms appear in chemical compounds by assigning different symbols to functional groups. Definitions of some of the WLN symbols involving oxygen atoms are as follows:

O   Oxygen, when connected to nonhydrogen atoms such as ethers or esters
Q   Hydroxyl group
V   Carbonyl group
W   Dioxo group as in $-NO_2$ or $-SO_2-$

Thus, oxygen atoms are represented by different symbols depending on the different possible environments.

For nitrogen, the following symbols are used:

Z   Nitrogen in $-NH_2$ (primary)
M   Nitrogen in $-NH-$ (secondary)
N   Nitrogen in $>N-$ (tertiary)

For carbon atoms, the following symbols are used:

C   Unbranched carbons doubly or triply bonded to at least one other element, such as nitrile

Numeral Straight-chain alkyl group

Y   Branch point carbon with three connections

X   Branch point carbon with four connections

The halogen atoms are coded as follows:

     F Fluorine
     E Chloride
     G Bromine
     I Iodine

Compounds are coded in WLN by choosing symbols and arranging them in the appropriate sequence following a set of rules. The three steps followed in generating the notation for a compound are:

1. Draw the two-dimensional structural diagram using WLN symbols.
2. Choose a path to trace through the structure.
3. Convert the drawing into the correct linear notation.

The conversion of the drawing into the correct linear notation is accomplished by following a set of rules. The rules build in complexity starting from the extremely simple to the complex. The following pages introduce some of the basic rules one by one with example structures. The complete rule set is evolving to cover all chemical compounds (Smith 1968).

Rule 1. Cite all chains of structural units symbol by symbol as connected.
 With only this rule and the symbols introduced above, we can generate the WLN for simple, symmetrical structures.

$$CH_3-\overset{\displaystyle \overset{O}{\|}}{C}-CH_3 \qquad 1V1$$

$$CH_3-CH_2-O-CH_2-CH_3 \qquad 2O2$$

$$HO-CH_2-CH_2-CH_2-OH \qquad Q3Q$$

$$H_2N-CH_2-CH_2-NH_2 \qquad Z2Z$$

Rule 2. To resolve precedence problems, select the sequence giving the notation with its starting symbol as late in the following sequence as possible.

$$b\&-/0123456789ABC\dots XYZ$$

$$CH_3-CH_2-\overset{\overset{\textstyle O}{\|}}{C}-CH_2-NH_2 \qquad\qquad ZIV2 \text{ (Z later than 2)}$$

$$HO-CH_2-CH_2-\overset{\overset{\textstyle O}{\|}}{C}-OH \qquad\qquad QV2Q \text{ (QV later than Q2)}$$

$$CH_3-CH_2-CH_2-NH-CH_2-CH_3 \qquad 3M2 \text{ (3 later than 2)}$$

Rule 3. Use U for a double bond in a chain when it is between:

a. Two atoms where cis/trans or syn/anti isomers are possible.
b. Carbon and a branching, divalent symbol.
c. Noncarbon and an alkyl numeral.
d. Carbon atoms in allenes and cumulenes.

The symbol UU denotes an acetylenic bond.

$$CH_3-CH=CH-CH=CH_2 \qquad 2U2U1$$

$$CH_3-CH_2-O-CH=NH \qquad MU1O2$$

$$CH_3-C\equiv C-CH_2-CH_3 \qquad 3UU2$$

$$CH_3-CH=C=N-CH_3 \qquad 2U1N1$$

$$CH_3-N=N-CH_2-CH_3 \qquad 2NUN1$$

Rule 4. Unsaturations not covered by rule 3 are denoted by directly joining the symbols of the atoms involved.

$$CH_3-CH_2-CH_2-C\equiv N \qquad NC3$$

$$CH_3-CH=C=O \qquad OC2$$

Rule 5. Use H only when not implied by another symbol. Use H in hydrocarbons.

$$CH_3-CH_2-CH_2-CH_3 \quad 4H$$

$$CH_3-CH_2-\overset{\overset{\displaystyle O}{\|}}{C}H \quad\quad VH2$$

For developing the WLN for branched compounds, several additional rules are necessary. These are rules 6, 7, and 8 in Smith (1968). The location in the structure where the notation is started is determined by a precedence sequence. In stepping through the molecule, when the path comes to a branch point, X or Y, use the following rules:

a. Cite the path with the fewest branch symbols, or
b. Cite the path with the fewest total symbols, or
c. Cite the path with the notation preferred by the precedence list.

The end of each path is shown by punctuating with an & unless the path ends with a normally terminal group such as a hydroxyl or a halogen.

In generating the WLN for branched compounds, it is helpful to draw out the structure using the WLN symbols, as in the following examples.

$$Cl-CH_2-CH_2-N\overset{\displaystyle CH_2-CH_3}{\underset{\displaystyle CH_2-CH_2-CH_3}{}}$$

G 2 N 2   G2N3&2
3

$$HO-CH_2-\overset{\overset{\displaystyle Cl}{|}}{\underset{\underset{\displaystyle Cl}{|}}{C}}-CH_2-CH_2-CH\overset{\displaystyle CH_2-NH_2}{\underset{\displaystyle CH_2-OH}{}}$$

G
Q1 X 2 Y 1Z
G 1
Q

Z1Y1Q2XGG1Q

$$\overset{\displaystyle CH_3-CH_2}{\underset{\displaystyle CH_3-CH_2}{}}CH-\overset{\overset{\displaystyle Cl}{|}}{C}H-\overset{\overset{\displaystyle O}{\|}}{C}-OH$$

2
2 Y Y VQ   QVYGY2&2
Z

$$\overset{\displaystyle CH_3-CH_2}{\underset{\displaystyle CH_3-CH_2-CH_2}{}}N-CH_2-\overset{\overset{\displaystyle CH_2Cl_2}{|}}{C}H-CH_2-\overset{\overset{\displaystyle CH_2OH}{|}}{\underset{\underset{\displaystyle CH_2CH_2OH}{|}}{C}}-F$$

G GQ
Y   1
2 N 1 Y 1 X F   Q2XF1Q1YYGG1N3&2
3   2
Q

Benzene rings are handled by representing them with the symbol R.

$$\langle\bigcirc\rangle-CH_2-CH_3 \qquad 2R$$

$$\langle\bigcirc\rangle-N\begin{array}{l}\diagup CH_2-CH_3\\ \diagdown CH_2-C-OH\\ \qquad\qquad\|\\ \qquad\qquad O\end{array} \qquad QV1N2\&R$$

The relative locations of substituents on the benzene ring are given by letters that follow blanks in the notation sequence. These space-followed-by-a-letter symbols are called locants. The locant A refers to the location on the ring where the paths being cited enter the ring, B is adjacent to this (ortho), C is two atoms away (meta), and D is three atoms away (para).

$$HO-\langle\bigcirc\rangle-CH_2-NH_2 \qquad Z1R\_DG$$

$$\begin{array}{l}HO\\ \langle\bigcirc\rangle-CH_2-NH_2 \qquad Z1R\_CG\end{array}$$

$$\begin{array}{l}OH\\ \langle\bigcirc\rangle-CH_2-NH_2 \qquad Z1R\_BG\end{array}$$

$$\begin{array}{l}Br\\ Cl-\langle\bigcirc\rangle-NH_2 \qquad ZR\_DG\_BE\end{array}$$

The order in which attachments to a multiply substituted benzene ring are cited is determined by the symbols representing the groups rather than by the locant letters.

$$\begin{array}{l}CH_3-CH_2\diagdown\\ \qquad\qquad N-\langle\bigcirc\rangle-\overset{\overset{\textstyle O}{\|}}{C}-NH-\langle\bigcirc\rangle-OCH_3\\ CH_3-CH_2\diagup\\ \qquad\qquad\qquad\qquad\qquad\qquad\qquad 2N2\&R\_CVMR\_DO1\end{array}$$

The handling of benzene rings is covered by rules 20–24 in Smith (1968).

Monocyclic ring systems other than benzene rings are divided into two classes: heterocyclic rings and carbocyclic rings. For heterocyclic rings, all the information relevant to the ring system is enclosed with a pair of symbols T and J, and for carbocyclic rings L and J are used. After citing the initial L or T, the following information is given:

1. A numeral giving the number of ring atoms.
2. Ring segment symbols.
3. Saturation symbols.
4. J to finish the ring description.

The symbols for ring substituents are cited in the alphabetic order of their locants. The saturation sign is T.

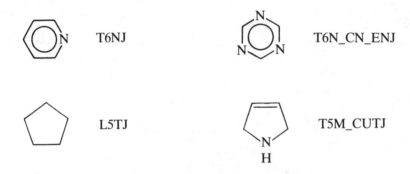

Rules have been generated for WLN to handle fused ring systems and other chemical species not covered in this introductory discussion. The full rule set contains several dozen rules of ever-increasing complexity and can be found in Smith (1968).

10.3 CONNECTION TABLE REPRESENTATION

The topology of molecular structures can be explicitly represented in the form of connection tables. To generate the connection table for a compound, the nonhydrogen atoms are numbered arbitrarily, and a table is constructed that cites all the interconnections. Table 1 shows the simple connection table for the seven-atom molecule, 3-aminobutanoic acid. In the usual case, the hydrogen atoms are not cited explicitly. It is obvious that each bond has been cited twice, which constitutes unnecessary and undesirable redundancy. This redundancy can be eliminated, without loss of information, by adopting the following two conventions:

TABLE 1. Connection Table for 3-Aminobutanoic Acid

		Connections					
Atom Number	Atom Type	Atom Number	Type	Atom Number	Type	Atom Number	Type
1	C	2	1	—	—	—	—
2	C	1	1	3	1	4	1
3	N	2	1	—	—	—	—
4	C	2	1	5	1	—	—
5	C	4	1	6	2	7	1
6	O	5	2	—	—	—	—
7	O	5	1	—	—	—	—

1. In numbering the structure, once an atom has been numbered, all unnumbered atoms connected to it are numbered serially.

2. In the connection table, only cite connections to lower numbered atoms.

Applying these rules to the same seven-atom molecule yields the compact connection table of Table 2. The numbering sequence used in Table 1 followed the numbering convention, so renumbering was not necessary in this example.

For cyclic structures, one extra line must be added to the compact connection table for each ring. This ring closure entry specifies the one extra bond found because of the presence of the ring.

Aromatic rings can be handled in several ways. Either the bond types can be entered as alternating single and double bonds or a special bond type can be adopted for aromatic bonds only.

Evidently, many numberings are possible for any given structure, and each different numbering will yield a different connection table. Thus, the connection tables generated by the procedure described above are not unique. However, they do completely specify the topology of the molecular structure, and therefore they are unambiguous.

Charges, abnormal valences, and so on, are stored as adjunct information in addition to the connection table.

The information stored in connection tables can be stored in an alternative mode utilizing binary matrices. Thus, the same information stored in the connection table shown in Table 2 could be stored as shown in Table 3.

As will be shown in a suceeding section, structures represented by

TABLE 2. Compact Connection Table for 3-Aminobutanoic Acid

Atom Number	Atom Type	Connections Atom Number	Type
1	C	—	—
2	C	1	1
3	N	2	1
4	C	2	1
5	C	4	1
6	O	5	2
7	O	5	1

TABLE 3. Binary Matrix Storage Mode for Connection Table of 3-Aminobutanoic Acid

Atom Connectivity Matrix

	1	2	3	4	5	6	7
1		1	0	0	0	0	0
2			1	1	0	0	0
3				0	0	0	0
4					1	0	0
5						1	1
6							0
7							

Atom Type Matrix

	1	2	3	4	5	6	7
C	1	1	0	1	1	0	0
O	0	0	0	0	0	1	1
N	0	0	1	0	0	0	0

Bond Type Matrix

	1	2	3	4	5	6
Single	1	1	1	1	0	1
Double	0	0	0	0	1	0

connectivity matrices, such as that shown in Table 3, are known as graphs in the sense used in the branch of mathematics known as mathematical graph theory. Thus, all the power, rigor, and theorems of graph theory can be applied to the study of chemical structures when they are represented as connection tables. This is one of the outstanding features that recommends the use of connection tables as a mode of chemical structure representation.

Program CTTEST

To illustrate the simplicity of the connection table mode of representation of chemical structures and to provide the means for entry of connection tables for later use, the program CTTEST and several subroutines are presented. Program CTTEST is a small driver program that is used here only to call two subroutines where the actual manipulations are done.

Subroutine CTIS (connection table input subroutine) is a minimal routine for the input of a connection table. In order to keep CTIS short and simple, no error checking or other usual features of such routines have been added. This routine contains only the essential elements necessary for input of a connection table. The user must input the number of atoms, the number of bonds, the atom types, and the bonding information. The atom types allowed and their codes are 1 for C, 2 for O, 3 for N, 4 for S, and 5 for Cl. The bonding information is entered as triplets containing the atom number from which the bond originates, the atom number where the bond terminates, and the bond type. Bond types of 1 for single, 2 for double, 3 for triple, and 4 for aromatic are supported. The information is stored in a named common block CTI (for connection table information). This common block allows communication between the subroutine CTIS and the routine that calls it.

Subroutine CTPR is a routine to print out the connection table on the interactive device for visual checking by the user. It outputs a connection table with atomic symbols on the main diagonal representing the atom types and numerals in the off-diagonal elements representing the bonds. Zeros are in the off-diagonal positions where no bonds are found.

An example input–output stream is shown for the methyl ester of propionic acid. This compound has six atoms, five bonds, four carbon atoms, two oxygen atoms, four single bonds, and one double bond. In the input–output stream, the messages written by the program start in column 2 or further over, and the user's input starts in column 1.

The subroutine CTIS will be used in other programs later in this book as a means for entering molecular structures for processing.

```
      PROGRAM CTTEST
C....
C.... CONNECTION TABLE INPUT AND OUTPUT TESTING PROGRAM
C....
C.... /CTI/ CONTAINS CONNECTION TABLE INFORMATION
C....      32 ATOMS & 32 BONDS MAXIMUM
      COMMON /CTI/ NATOMS,NBONDS,IFA(32),ITA(32),NBTYPE(32),NATYPE(32)
C.... /IOUNIT/ CONTAINS THE I/O UNIT NUMBERS
      COMMON /IOUNIT/ NINP,NOUT
C....
      NINP=1
      NOUT=1
      WRITE (NOUT,2)
    2 FORMAT (' CONNECTION TABLE TEST PROGRAM',/)
      CALL CTIS
      CALL CTPR
      STOP
      END
```

```
C-----------------------------------------------------------------
      SUBROUTINE CTIS
C....
C.... CONNECTION TABLE INPUT SUBROUTINE
C....
      COMMON /CTI/ NATOMS,NBONDS,IFA(32),ITA(32),NBTYPE(32),NATYPE(32)
      COMMON /IOUNIT/ NINP,NOUT
C....
      WRITE (NOUT,1)
    1 FORMAT (' CONNECTION TABLE INPUT',/)
      WRITE (NOUT,2)
    2 FORMAT (' INPUT NATOMS, NBONDS')
C.... INPUT NUMBER OF ATOMS AND NUMBER OF BONDS
      READ (NINP,*) NATOMS,NBONDS
      WRITE (NOUT,3)
    3 FORMAT (' INPUT ATOMS TYPES: 1=C, 2=O, 3=N, 4=S, 5=CL')
C.... INPUT ATOM TYPES
      READ (NINP,*) (NATYPE(K),K=1,NATOMS)
      WRITE (NOUT,4)
    4 FORMAT (' INPUT TRIPLES: (FROM ATOM, TO ATOM, BOND TYPE)')
C.... INPUT BONDING INFORMATION
      READ (NINP,*) (IFA(K),ITA(K),NBTYPE(K),K=1,NBONDS)
      RETURN
      END


C-----------------------------------------------------------------
      SUBROUTINE CTPR
C....
C.... SUBROUTINE TO PRINT OUT CONNECTION TABLE
C....
      INTEGER*2 IARRYA(5),IARRYB(4),KCT(32,32)
      COMMON /CTI/ NATOMS,NBONDS,IFA(32),ITA(32),NBTYPE(32),NATYPE(32)
      COMMON /IOUNIT/ NINP,NOUT
      DATA IARRYA /' C',' O',' N',' S','CL'/
      DATA IARRYB /' 1',' 2',' 3',' 4'/
      DATA IBLANK /' '/
C....
      WRITE (NOUT,1)
    1 FORMAT (' ',/,' CONNECTION TABLE OUTPUT',/)
      DO 20 I=1,NATOMS
      DO 20 J=1,NATOMS
   20 KCT(I,J)=IBLANK
      DO 30 I=1,NBONDS
      KCT(IFA(I),ITA(I))=IARRYB(NBTYPE(I))
   30 KCT(ITA(I),IFA(I))=IARRYB(NBTYPE(I))
      DO 40 I=1,NATOMS
   40 KCT(I,I) = IARRYA(NATYPE(I))
      WRITE (NOUT,42) (I,I=1,NATOMS)
   42 FORMAT (' ',3X,32(I2),/)
      DO 45 I=1,NATOMS
   45 WRITE (NOUT,47) I,(KCT(I,J),J=1,NATOMS)
   47 FORMAT (' ',I2,1X,32(A2))
      RETURN
      END


      CONNECTION TABLE TEST PROGRAM

      CONNECTION TABLE INPUT

      INPUT NATOMS, NBONDS
      6 5
      INPUT ATOMS TYPES: 1=C, 2=O, 3=N, 4=S, 5=CL
      1 1 1 2 2 1
      INPUT TRIPLES: (FROM ATOM, TO ATOM, BOND TYPE)
      1 2 1 2 3 1 3 4 2 3 5 1 5 6 1
```

```
CONNECTION TABLE OUTPUT

    1 2 3 4 5 6
 1   C 1
 2   1 C 1
 3     1 C 2 1
 4       2 0
 5       1   0 1
 6             1 C
```

REFERENCES

G. W. Gibson and C. E. Granito, "The Wiswesser Line-Formula Notation," *Amer. Lab.*, **4**, 27–37 (1972).

M. F. Lynch, J. M. Harrison, W. G. Town, *Computer Handling of Chemical Structure Information*, Macdonald, London, 1971.

E. G. Smith, *The Wiswesser Line-Formula Chemical Notation*, McGraw-Hill, New York, 1968.

W. J. Wiswesser, "Historic Development of Chemical Notations," *J. Chem. Inf. Comput. Sci.*, **25**, 258–263 (1985).

J. E. Ash et al., *Communication Storage and Retrieval of Chemical Information*, Halsted Press, New York, 1985.

11

MATHEMATICAL GRAPH THEORY

11.1 INTRODUCTION

Graph theory is the study of the nature and properties of topological graphs. It is a discipline within mathematics that has a multitude of applications to chemistry. It was first developed in the 1700s and has found application throughout science and engineering and beyond.

In the sense used here, the term *graph* must not be confused with the familiar Cartesian coordinate plots of experimental data. In graph theory, a graph is an abstract concept applied to a collection of points (also called nodes or vertices) joined by lines (also called edges). Edges connect pairs of vertices together. Some examples of graphs are shown in Figure 1.

Some of the fundamental definitions of graph theory follow. A graph is a finite collection of points (also called vertices or nodes) and a finite collection of lines (also called edges). The edges connect pairs of vertices together. Two vertices connected by an edge are *adjacent*. A *walk* of a graph is an alternating sequence of points and lines that begins and ends with points. The walk becomes a *path* if all the points traversed are distinct. The walk is a *cycle* if it is closed and if at least three distinct points are traversed. The *length* of a walk is the number of lines in it. The *distance* between two points in a graph is the length of the shortest path joining them. The *degree* of a point is the number of lines connected to it. A graph is *connected* if every pair of points are joined by a path. A graph is *acyclic* if it has no cycles. A *tree* is a connected, acyclic graph. A graph is *labeled* when its nodes are distinguishable from one another by names.

An obvious use for graph theory is for depicting molecular structure. The points of the graph are the atoms of the molecule, and the edges of the graph are the bonds of the molecule. The degree of an atom is the number of bonds connected to it, the valency.

156

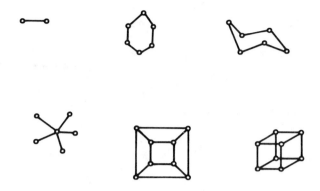

Figure 1. Examples of graphs.

Figure 2 shows all 23 trees that have 8 nodes. Five of these are chemically implausible (as carbon-backbone compounds) because they contain nodes of degree greater than 4. Thus, if 4 is the maximum degree allowed, there are exactly 18 possible tree structures with 8 nodes.

The structure of a graph can be represented by an adjacency matrix, $A = a_{ij}$, where $a_{ij} = 1$ if nodes i and j are adjacent and 0 otherwise. This is the same as the binary atom connectivity matrix introduced in Chapter 10. Thus, a link exists between the representation of chemical structures as connection tables and mathematical graph theory. Theorems developed in the abstract domain of graph theory can be applied directly to chemical structures in many cases. For example, the row sums of the adjacency matrix are the connectivities of the atoms. The number of paths of length n within a molecule can be found by calculating A^n. The value for the i, j entry in the resulting matrix is the number of paths of length n from atom i to atom j.

Once the connection between graph theory and chemical structure representation has been established, the theorems of graph theory can be exploited for chemical purposes. This has led to important advances in a number of areas of chemistry, including the following:

1. Registration of chemical compounds.
2. Isomer enumeration.
3. Ring locating.
4. Substructure searching.
5. Quantification of structural similarity.

Many reviews of the applications of graph theory to chemical problems have appeared in the literature (e.g., Balaban 1976, 1985).

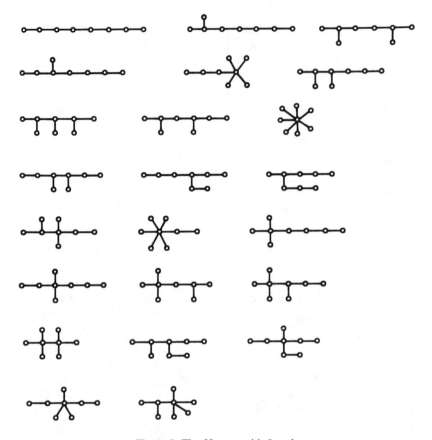

Figure 2. The 23 trees with 8 nodes.

11.2   REGISTRATION OF CHEMICAL COMPOUNDS

The ability to store individual chemical compounds in a computer-compatible form using a linear notation or connection table representation leads to the ability to generate files of compounds. In many industrial companies, such compound files contain tens to hundreds of thousands of structures, often with associated data. In the process of building up such files, upon receipt of each compound to be added to the file, it is necessary to answer the question "Is this compound already in this file?" If the answer is yes, then the compound need not be added. If the answer is no, a new entry is necessary. This process is called *registration*. The ability to perform the registration function accurately and efficiently is crucial in systems such as the Chemical Abstracts Service registry system. In this system, information relating to a compound is stored under the compound's identifying name.

If the compound is represented by a linear notation, answering this registration question amounts to doing a search through the file because the notation is unique. If the compound were already in the file, it would have been represented by precisely the same linear notation when originally entered. Therefore, searching the files for the query compound amounts to looking for the same string of symbols that represent the query compound in the already existent file.

However, as we have seen in Chapter 10, a connection table representation of a chemical structure is not unique. To make it unique, which greatly simplifies the registration problem, algorithms based on graph theory have been developed. The most important one is known as the Morgan algorithm (Morgan 1965) after its developer.

A given chemical compound can be represented by many equivalent connection tables, depending on how the numbers were assigned to the atoms. Generating a unique connection table, a *canonical* connection table, is done by generating a unique, invariant numbering for the structure based on its own structure. If any arbitrary numbering sequence is allowed, there are $n!$ (n factorial) numbering schemes for a molecule with n nonhydrogen atoms. This presents a gargantuan number of potential numbering sequences for molecules of even moderate size. The potential number of sequences can be reduced dramatically by following the numbering rules cited in Section 10.3. That is, once an atom has been numbered, then all unnumbered atoms connected to it are numbered serially. While the number of sequences that can arise from this scheme is not easily calculated, it is still extremely large. There must be a method for choosing just one numbering sequence in a repeatable way. The Morgan algorithm utilizes the structure of the compound being treated to resolve this problem. It focuses on the extended connectivities of the atoms (the number of nonhydrogen atoms connected to them) and the atom types to sort the atoms. Expressed in words, the algorithm seeks the most deeply imbedded atom within the structure to give the number 1. Then it follows the serial numbering rule, resolving ties by the extended connectivities of the atoms involved, until the structure is completely numbered. In algorithm form, the procedure is as follows (O'Korn 1977):

1. Calculate stage 1 connectivity values for each atom by inspecting the structure.

2. Calculate stage 2 connectivity values for each atom by summing the stage 1 connectivity values for the attached atoms.

3. Calculate stage $i + 1$ connectivity values for each atom by summing the stage i connectivity values for the attached atoms.

4. At each stage, calculate the number of distinct connectivity values that appear in the structure, k.

5. As long as the number of distinct connectivity values continues to increase, keep repeating steps 3 and 4.

6. When a stage is found where the number of distinct connectivity values stays equal or decreases, save the connectivity values from the previous stage.

7. The atom with the highest connectivity value is given the number 1.

8. Note all other atoms with the same connectivity value for later use.

9. Serially number all atoms connected to atom 1 as number 2, 3, and so on, based on decreasing connectivity values. If an arbitrary choice must made, note the pairs of atoms involved in the arbitrary choice for later use.

10. The unnumbered atoms connected to atom 2 are serially numbered based on decreasing connectivity values. As before, note the pairs of atoms involved in any arbitrary choices.

11. Follow this procedure until all the atoms have numbers.

12. Generate the compact connection table based on this numbering sequence. If no arbitrary choices were made during the numbering, this connection table is the canonical one, and the procedure is terminated.

13. If there were arbitrary choices made, back up to the highest numbered atom for which an arbitrary choice was made.

14. Select the other atom from the pair involved in the arbitrary choice, and renumber the atoms of the structure from that atom to the last atom.

15. Generate the compact connection table based on this numbering sequence.

16. Compare the newly generated connection table to the previously generated one.

17. If the new connection table is better than the previous one, replace the retained one with the new one.

18. Go back to step 13.

In step 17 the idea of better was introduced. To generate canonical numbering, it is not necessary for the definition of *better* to be sophisticated, only invariant. In the original formulation, alphabetic ordering was used. This is equivalent to using atom types to decide precedence.

In Figure 3, the application of the Morgan algorithm to a nine-atom molecule, 2-isopropyl-4-methyltetrahydrofuran, is shown. The initial set of connectivities contains the three values $(1, 2, 3)$ and $k = 3$. By stage 3, k has increased to 6, and it does not increase further on, going to stage 4. Therefore the stage 3 connectivity values are taken as the final ones for use during numbering. During numbering, an arbitrary choice must be made when assigning the number 3. Therefore, two different sequences have been generated, and two corresponding compact connection tables can be generated. The arbitrary choice in assigning the number 5 is unimportant because atoms 5 and 6 are identical through symmetry. The two compact connection tables are compared to determine which is better, and the one corresponding to the oxygen atom numbered 4 is chosen because it is desirable to have heteroatoms appear late in the connection table. The actual canonical "name" of this compound to be used in the registration

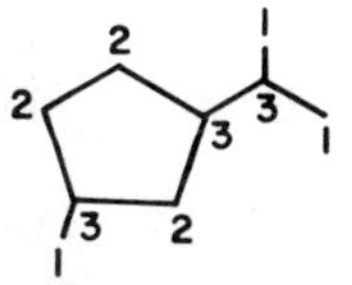

Stage 1. [1, 2, 3] $k = 3$

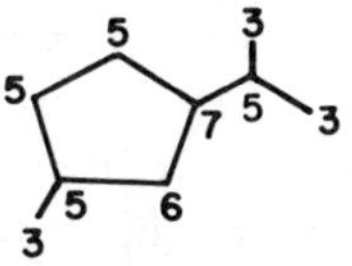

Stage 2. [3, 5, 6, 7] $k = 4$

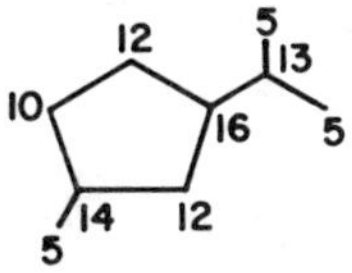

Stage 3. [5, 10, 12, 13, 14, 16] $k = 6$

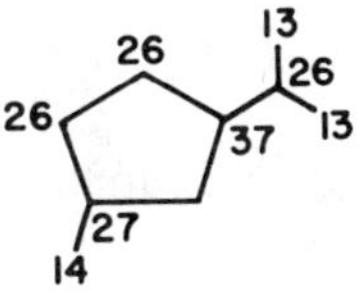

Stage 4. [13, 14, 26, 27, 30, 37] $k = 6$

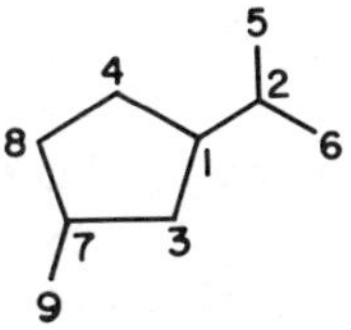

Numbering system resulting from arbitrary choice in assignment of 3

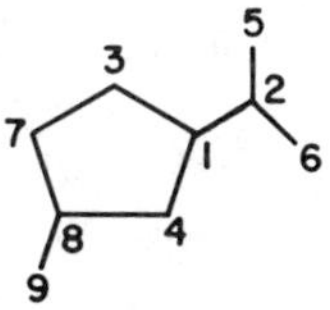

Numbering system resulting from arbitrary choice in assignment of 3

Compact Connection Table

Atom Number	Atom Type	Connections: Atom Number	Connections: Type
1	C	—	—
2	C	1	1
3	C	1	1
4	O	1	1
5	C	2	1
6	C	2	1
7	C	3	1
8	C	4	1
9	C	7	1

Ring closure (7, 8)

Figure 3. Application of the Morgan algorithm to 2-isopropyl-4-methyltetrahydrofuran.

process is then formed from this preferred compact connection table. The string of symbols formed by sweeping out the connection table is unique and can be used directly in searching files for a match.

A number of extensions of this basic algorithm have been developed to deal with additional complexities of molecular structures. For example, the stereochemistry of structures can be handled by using extensions described by Wipke and Dyott (1974).

11.3 ENUMERATION OF ISOMERS

The first work on isomer enumeration using graph theory was by Cayley in 1874, and it dealt with tree structures. Techniques involving recursion formulas were first developed by Henze and Blair (1932). This approach involves using the isomer count for a given number of homologous series containing n carbon atoms and then computing the isomer count for the succeeding member containing $n + 1$ carbon atoms. Many authors have applied this recursive approach to a variety of chemical systems (Rouvray 1974). These methods are particularly well suited to computer implementation.

The single most important advance in isomer enumeration was Polya's enumeration theorem (Polya 1937). It exploited group theory. It is presented and discussed in many secondary sources (e.g., Harary 1969, Harary et al. 1976).

The field of isomer enumeration continues to attract the attention of mathematical chemists. Lederberg and co-workers developed a general procedure for enumerating the acyclic isomers corresponding to a given elemental composition (Lederberg et al. 1969), and the algorithm, named DENDRAL and later CONGEN, has been improved several times. Recently, this group has reported an algorithm capable of generating not only constitutional isomers but stereoisomers as well (Nourse et al. 1980).

11.4 ATOMIC AND MOLECULAR PATH COUNTS

The use of graph theoretic properties of chemicals represented as graphs for chemical purposes has attracted a great deal of attention. Another example is that of atomic and molecular paths. In the context of graph theory, a *path* is defined as a sequence of nodes in a graph that are connected by edges, where no node appears more than once in the list. The *length* of a path is the number of edges in the path. Since molecules can be represented as graphs, their paths can be investigated as a chemical property. The enumeration of the paths in a molecule becomes progressively more difficult as the molecule increases in size, in innerconnectivity, or both.

To illustrate the idea of paths within molecules, consider the example molecule shown in Figure 4. The molecule is drawn in the usual two-

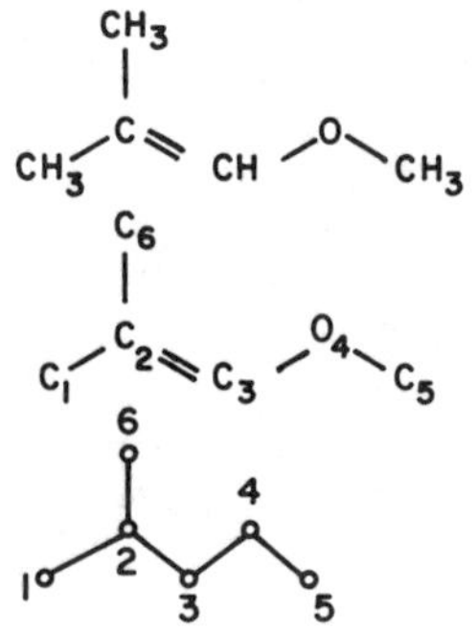

Atom Number	Path Length				
	1	2	3	4	5
1	1	2	1	1	0
2	3	1	1	0	0
3	2	3	0	0	0
4	2	1	2	0	0
5	1	1	1	2	0
6	1	2	1	1	0
Molecular Path Count	5	5	3	2	0

Figure 4. Path lengths for example molecule.

dimensional structural diagram, then with the hydrogen atoms suppressed and the atoms numbered, and then as a numbered graph.

Focus first on atom 1. There is only one path of length 1, namely 1–2. There are two paths of length 2, namely 1–2–3 and 1–2–6. There is one path of length 3 and one path of length 4. For atom 1 of the example molecule, the atom path code is 1, 2, 1, 1. For the other atoms in the structure, the atom path codes are different, according to the degree to which they participate in paths of different lengths. Figure 5 shows the atom path codes for all the isomers of hexane.

In acyclic structures, there is a unique path between any pair of atoms, so the number of paths of a specified length gives the number of neighbors at that distance. The number of neighbors a given distance away from the atom of interest is the basis for several schemes for ^{13}C NMR shift predictions by additivity methods (e.g., Grant and Paul 1964, Lindeman and Adams 1971).

Figure 5 shows the atom path codes for all five isomers of hexane. Each of these isomers has a total of 15 unique paths, but they are distributed differently among the possible lengths. Some regularities are evident in the codes for these structures. For example, the length of the longest path for each atom in a structure is the number of bonds that must be traversed to go from that atom to the most distant terminal atom. Only a completely extended alkane of n atoms with no internal branching can have a path of length $n - 1$. There is a general relationship between the "extent" of a molecule and the length of the paths found within it.

Another representation based on paths, the *molecular path code*, is obtained by summing the atomic path codes by lengths. Each path appears twice, once for each end atom, so the results must be divided by 2. Generally, different molecules have different molecular path codes. The molecular path code for each of the five hexane isomers is shown in Figure 5. The average path length for each isomer is also shown. The most extended isomer, n-hexane, has the largest average value of 2.33, and the most highly branched isomer, 2,2-dimethylbutane, has the smallest average path length of 1.87.

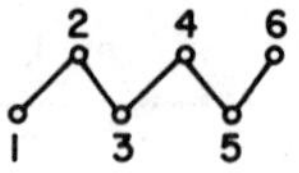

1	1, 1, 1, 1, 1
2	2, 1, 1, 1
3	2, 2, 1
4	2, 2, 1
5	2, 1, 1, 1
6	1, 1, 1, 1, 1

5, 4, 3, 2, 1   mean = 2.33

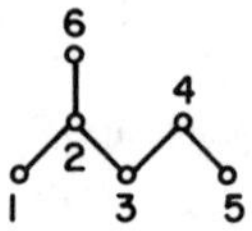

1	1, 2, 1, 1
2	3, 1, 1
3	2, 3
4	2, 1, 2
5	1, 1, 1, 2
6	1, 2, 1, 1

5, 5, 3, 2   mean = 2.13

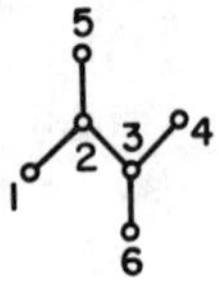

1	1, 2, 2
2	3, 2
3	3, 2
4	1, 2, 2
5	1, 2, 2
6	1, 2, 2

5, 6, 4   mean = 1.93

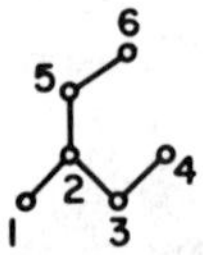

1	1, 2, 2
2	3, 2
3	2, 2, 1
4	1, 1, 2, 1
5	2, 2, 1
6	1, 1, 2, 1

5, 5, 4, 1   mean = 2.00

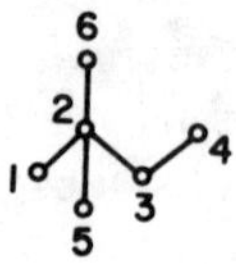

1	1, 3, 1
2	4, 1
3	2, 3
4	1, 1, 3
5	1, 3, 1
6	1, 3, 1

5, 7, 3   mean = 1.87

Figure 5. The atom path codes for all five isomers of hexane.

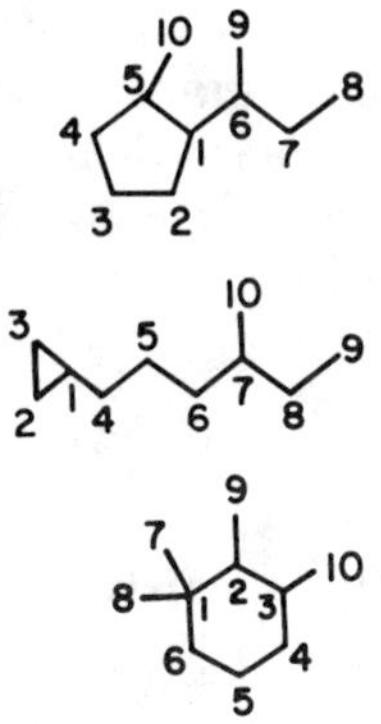

| Compound Q | 10 | 13 | 16 | 17 | 11 | 7 | 4 | 1 |

| Compound A | 10 | 12 | 10 | 8 | 7 | 7 | 6 | 2 |

| Compound Y | 10 | 15 | 17 | 16 | 14 | 10 | 3 |

$(QA) = 139$
$(QY) = \ \ 26$
$(AY) = 193$

Figure 6. Dissimilarities among three monocyclic monoterpenes.

Molecular path codes provide a method for studying similarity between molecules quantitatively (Randic and Wilkins 1979). They computed the structural similarities among a set of 29 monocyclic monoterpenes by evaluating the following dissimilarity measure:

$$D^2_{ab} = \sum_i (a_i - b_i)^2$$

where a and b refer to the molecular path codes for two different structures, and the summation is carried out over the index i, which indicates the lengths of the paths. Here $D = 0$ for molecules with identical molecular path codes, and it grows for less similar structures. Figure 6 shows the computation of the dissimilarities among three monocyclic monoterpenes. Compounds A and Y are quite dissimilar while compounds G and Y are quite similar. For all $\frac{1}{2}(29 \times 28) = 406$ comparisons, the smallest value of D reported was 3 and the largest was 306. Randic and Wilkins discussed possible uses of this dissimilarity measure based on molecular path counts in the areas of structure elucidation and structure chemistry.

Program PATH

Program PATH is a FORTRAN language routine that enumerates the atomic and molecular paths of a molecule represented as a connection table. The computation is done by subroutine ALPATH, which is a somewhat revised version of the routine published by Randic and co-workers (1979). The structure being considered is input through subroutine CTIS and is passed to ALPATH through the common block named CTI. The quantity of output generated is determined by the user's answer to a query. If less output is desired, the numbers of paths of each length for each atom are output. If more output is specified, the detailed enumeration of each and every path in the structure is also output. Several examples of the output generated for molecules of varying complexity are given in the accompanying figures.*

```
        PROGRAM PATH
C. . . .
C. . . . PROGRAM TO ENUMERATE ATOMIC AND MOLECULAR PATHS
C. . . .
        DIMENSION LB(32),LC(32,32)
        COMMON /CTI/ NATOMS,NBONDS,IFA(32),ITA(32),NBTYPE(32),NATYPE(32)
        COMMON /IOUNIT/ NINP,NOUT
        DATA IYES/'Y'/
        NINP=1
        NOUT=1
        WRITE (NOUT,1)
      1 FORMAT (' PATH PROGRAM',/)
C. . . .
        WRITE (NOUT,3)
      3 FORMAT (' ENTER OUTPUT LEVEL: ',/,5X,'1=LEAST,  2=MOST')
        READ (NINP,*) IPR
C. . . .
     12 CALL CTIS
C. . . .
        DO 25 I=1,NATOMS
        DO 25 J=1,NATOMS
     25 LC(I,J)=0
        DO 27 J=1,NBONDS
        LC(IFA(J),ITA(J)) = 1
     27 LC(ITA(J),IFA(J)) = 1
        NSIZE=NATOMS
        CALL ALPATH (LB,LC,NSIZE,ITOT,IPR)
C. . . .
        WRITE (NOUT,55)
     55 FORMAT (' ANOTHER MOLECULE ? (Y OR N)')
        READ (NINP,56) IANS
     56 FORMAT (A1)
        IF (IANS.EQ.IYES) GO TO 12
        STOP
        END
C------------------------------------------------------------
        SUBROUTINE ALPATH (B,C,SIZE,TOTAL,PRFLG)
C. . . .
C. . . . M. RANDIC, G.M. BRISSEY, R.B. SPENCER, C.L. WILKINS,
C. . . .     COMPUTERS AND CHEMISTRY, 3, 5-13 (1979).
C. . . .
        IMPLICIT INTEGER (A-Z)
        DIMENSION A(32),B(32),C(32,32),R(32),S(32),P(32),BUFFER(32)
        COMMON /IOUNIT/ NINP,NOUT
C. . . . A ARRAY IS ATOM-BASED PATH COUNT VECTOR
C. . . . B ARRAY IS MOLECULE-BASED PATH COUNT VECTOR
C. . . . C ARRAY IS CONNECTION MATRIX
C. . . . R(I) INDICATES PREVIOUS ATOM IN CURRENT PATH
C. . . . S(I) INDICATES CURRENT COLUMN IN CONDENSED C CATRIX FOR ATOM I
C. . . . TOTAL RETURNS TOTAL NUMBER OF PATHS OF ALL LENGTHS
C. . . . PRFLG CONTROLS OUTPUT LEVEL -- 2 FOR SPECIFICATION OF ALL PATHS,
C. . . .      1 FOR ATOMIC PATH COUNTS, 0 FOR NONE
        DATA P/' 1',' 2',' 3',' 4',' 5',' 6',' 7',' 8',' 9','10','11',
       X  '12','13','14','15','16','17','18','19','20','21','22','23',
       X  '24','25','26','27','28','29','30','31','32'/
        DATA BLANK,BUFFER/33*'  '/,DIM/32/
```

* Subroutine ALPATH is reprinted with permission from *Computers and Chemistry*, **3**, 5–13 (1979). Copyright 1979, Pergamon Press, Ltd.

```
      IF (PRFLG. GT. 0) WRITE (NOUT, 1000)
      DO 340 I=1, SIZE
      K=1
      DO 330 J=1, SIZE
      IF (C(I, J). EQ. 0) GO TO 330
      C(I, K)=J
      IF (J. EQ. K) GO TO 320
      C(I, J)=0
320   K=K+1
330   CONTINUE
340   CONTINUE
      DO 344 J=1, DIM
344   B(J)=0
      DO 800 SATOM=1, SIZE
      IF (PRFLG. NE. 0) WRITE (NOUT, 1070) SATOM
      DO 345 J=1, DIM
      A(J)=0
      S(J)=0
345   R(J)=0
      N=1
      BUFFER(N)=P(SATOM)
      R(SATOM)=-1
      A(1)=1
      K=SATOM
480   S(K)=S(K)+1
      T=S(K)
      T=C(K, T)
      IF (T. EQ. 0) GO TO 600
      IF (R(T). NE. 0) GO TO 480
      R(T)=K
      K=T
      N=N+1
      IF (PRFLG. EQ. 0) GO TO 560
      BUFFER(N)=P(K)
560   BUFFLG=1
      A(N)=A(N)+1
      GO TO 480
600   IF (BUFFLG. EQ. 0) GO TO 640
      IF (PRFLG. LT. 2) GO TO 630
      WRITE (NOUT, 1030) (BUFFER(NQ), NQ=1, N)
630   BUFFLG=0
640   DO 650 IQ=1, N
650   BUFFER(IQ)=BLANK
      N=N-1
      IF (N. LT. 0) GO TO 720
      S(K)=0
      T=R(K)
      R(K)=0
      K=T
      IF (K. GT. 0) GO TO 480
720   DO 730 IQ=1, SIZE
730   B(IQ)=B(IQ)+A(IQ)
      IF (PRFLG. NE. 0) WRITE (NOUT, 1040) (A(IQ), IQ=1, SIZE)
800   CONTINUE
      TOTAL=B(1)
      DO 820 IQ=2, SIZE
      B(IQ)=B(IQ)/2
820   TOTAL=TOTAL+B(IQ)
      IF (PRFLG. EQ. 0) GO TO 821
      WRITE (NOUT, 1050) (B(IQ), IQ=1, SIZE)
      WRITE (NOUT, 1090) TOTAL
821   RETURN
1000  FORMAT (/, ' PATH ENUMERATION ROUTINE', /)
1030  FORMAT (' ', 25('    ', A2))
1040  FORMAT (' ', 2OI5)
1050  FORMAT (/, ' MOLECULAR PATH COUNTS', /, ' ', 2SI5)
1070  FORMAT (' PATHS FOR ATOM', I3)
1090  FORMAT (/, ' NUMBER OF PATHS =', I7)
      END
```

```
PATH PROGRAM

ENTER OUTPUT LEVEL:
    1 = LEAST, 2 = MOST
1
CONNECTION TABLE INPUT

INPUT NATOMS, NBONDS
6 5
INPUT ATOMS TYPES: 1=C, 2=O, 3=N, 4=S, 5=CL
1 1 1 2 1 1
INPUT TRIPLES: (FROM ATOM, TO ATOM, BOND TYPE)
1 2 1 2 3 2 2 6 1 3 4 1 4 5 1

PATH ENUMERATION ROUTINE

PATHS FOR ATOM  1
    1    1    2    1    1    0
PATHS FOR ATOM  2
    1    3    1    1    0    0
PATHS FOR ATOM  3
    1    2    3    0    0    0
PATHS FOR ATOM  4
    1    2    1    2    0    0
PATHS FOR ATOM  5
    1    1    1    1    2    0
PATHS FOR ATOM  6
    1    1    2    1    1    0


MOLECULAR PATH COUNTS
    6    5    5    3    2    0

NUMBER OF PATHS =     21
ANOTHER MOLECULE ? (Y OR N)
N

PATH PROGRAM

ENTER OUTPUT LEVEL:
    1 = LEAST, 2 = MOST
1
CONNECTION TABLE INPUT

INPUT NATOMS, NBONDS
8 12
INPUT ATOMS TYPES: 1=C, 2=O, 3=N, 4=S, 5=CL
1 1 1 1 1 1 1 1
INPUT TRIPLES: (FROM ATOM, TO ATOM, BOND TYPE)
1 2 1 1 4 1 1 5 1
2 3 1 2 6 1
3 4 1 3 7 1
4 8 1
5 6 1 5 8 1
6 7 1
7 8 1

PATH ENUMERATION ROUTINE

PATHS FOR ATOM  1
    1    3    6   12   18   30   24   18
PATHS FOR ATOM  2
    1    3    6   12   18   30   24   18
PATHS FOR ATOM  3
    1    3    6   12   18   30   24   18
PATHS FOR ATOM  4
    1    3    6   12   18   30   24   18
PATHS FOR ATOM  5
    1    3    6   12   18   30   24   18
PATHS FOR ATOM  6
    1    3    6   12   18   30   24   18
```

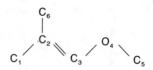

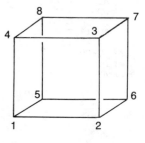

```
PATHS FOR ATOM  7
    1    3    6   12   18   30   24   18
PATHS FOR ATOM  8
    1    3    6   12   18   30   24   18

MOLECULAR PATH COUNTS
    8   12   24   48   72  120   96   72

NUMBER OF PATHS =    452
ANOTHER MOLECULE ? (Y OR N)
N

PATH PROGRAM

ENTER OUTPUT LEVEL:
    1 = LEAST,  2 = MOST
1
CONNECTION TABLE INPUT

INPUT NATOMS, NBONDS
8 7
INPUT ATOMS TYPES: 1=C,  2=O,  3=N,  4=S,  5=CL
1 1 1 1 1 1 1 1
INPUT TRIPLES: (FROM ATOM,  TO ATOM,  BOND TYPE)
1 2 1 2 3 1 3 4 1 4 5 1 5 6 1 6 7 1 7 8 1

PATH ENUMERATION ROUTINE

PATHS FOR ATOM  1
    1    1    1    1    1    1    1    1
PATHS FOR ATOM  2
    1    2    1    1    1    1    1    0
PATHS FOR ATOM  3
    1    2    2    1    1    1    0    0
PATHS FOR ATOM  4
    1    2    2    2    1    0    0    0
PATHS FOR ATOM  5
    1    2    2    2    1    0    0    0
PATHS FOR ATOM  6
    1    2    2    1    1    1    0    0
PATHS FOR ATOM  7
    1    2    1    1    1    1    1    0
PATHS FOR ATOM  8
    1    1    1    1    1    1    1    1

MOLECULAR PATH COUNTS
    8    7    6    5    4    3    2    1

NUMBER OF PATHS =     36
ANOTHER MOLECULE ? (Y OR N)
N
```

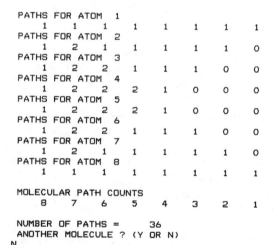

REFERENCES

A. T. Balaban, "Applications of Graph Theory in Chemistry," *J. Chem. Inf. Comput. Sci.*, **25**, 334–343 (1985).

A. T. Balaban (Ed.), *Chemical Applications of Graph Theory*, Academic, New York, 1976.

A. Cayley, *Philos. Mag.*, **67**, 444 (1874).

J. W. Essam and M. E. Fisher, "Some Basic Definitions in Graph Theory," *Rev. Mod. Phys.*, **42**, 272–288 (1970).

D. M. Grant and E. G. Paul, "Carbon-13 Magnetic Resonance. II. Chemical Shift Data for the Alkanes," *J. Amer. Chem. Soc.*, **86**, 2984–2990 (1964).

F. Harary, *Graph Theory*, Addison-Wesley, Reading, MA, 1969.

F. Harary, E. M. Palmer, R. W. Robinson, and R. C. Read, "Polya's Contributions to Chemical Enumeration," in *Chemical Applications of Graph Theory*, A. T. Balaban (Ed.), Academic, New York, 1976.

H. R. Henze and C. M. Blair, *J. Amer. Chem. Soc.*, **53**, 3042 (1931); **53**, 3077 (1931); **54**, 1098 (1932); **54**, 1538 (1932).

J. W. Kennedy, "Small Graphs, Graph Theory and Chemistry," in *Data Processing in Chemistry*, Z. Hippe (Ed.), Elsevier Scientific, Amsterdam, 1981.

J. Lederberg, G. L. Sutherland, and B. G. Buchanan, E. A. Feigenbaum, A. V. Robertson, A. M. Duffield, and C. Djerassi, "Applications of Artificial Intelligence for Chemical Inference. I. The Numbers of Possible Organic Compounds, Acyclic Structures Containing C, H, O, and N." *J. Amer. Chem. Soc.*, **91**, 2973–2976 (1969).

L. P. Lindeman and J. Q. Adams, "Carbon-13 Nuclear Magnetic Resonance Spectroscopy. Chemical Shifts for the Paraffins through C9," *Anal. Chem.* **43**, 1245–1252 (1971).

M. F. Lynch, J. M. Harrison, and W. G. Town, *Computer Handling of Chemical Structure Information*, Macdonald, London, 1971.

H. L. Morgan, "Generation of a Unique Machine Description for Chemical Structures: A Technique Developed at Chemical Abstracts Service," *J. Chem. Soc.*, **5**, 107–113 (1965).

J. G. Nourse, D. H. Smith, R. E. Carhart, and C. Djerassi, "Applications of Artificial Intelligence for Chemical Inference. 33. Computer-Assisted Elucidation of Molecular Structure with Stereochemistry," *J. Amer. Chem. Soc.*, **102**, 6289–6295 (1980).

L. J. O'Korn, "Algorithms in the Computer Handling of Chemical Information," in *Algorithms for Chemical Computations*, R. E. Christoffersen (Ed.), American Chemical Society, Washington, DC, 1977.

G. Polya, "Kombinatorische Anzahlbestimmungen fur Gruppen, Graphen und Chemische Verbindungen," *Acta Math.*, **68**, 145–254 (1937).

M. Randic, "Characterization of Atoms, Molecules, and Classes of Molecules Based on Path Enumerations," *MATCH*, **7**, 5–64 (1979).

M. Randic and C. L. Wilkins, "Graph Theoretical Approach to Recognition of Structural Similarity in Molecules," *J. Chem. Inf. Comput. Sci.*, **19**, 31–37 (1979).

M. Randic, G. M. Brissey, R. B. Spencer, and C. L. Wilkins, "Search for All Self-Avoiding Paths for Molecular Graphs," *Comput. Chem.*, **3**, 5–13 (1979).

M. Randic, G. M. Brissey, R. B. Spencer, and C. L. Wilkins, "Use of Self-Avoiding Paths for Characterization of Molecular Graphs with Multiple Bonds," *Comput. Chem.* **4**, 27–43 (1980).

D. H. Rouvray, "Isomer Enumeration Methods," *Chem. Soc. Rev.*, **3**, 355–372 (1974).

W. T. Wipke, S. R. Heller, R. J. Feldmann, and E. Hyde (Eds.), *Computer Representation and Manipulation of Chemical Information*, Wiley-Interscience, New York, 1974.

W. T. Wipke and T. M. Dyott, "Stereochemically Unique Naming Algorithm," *J. Amer. Chem. Soc.*, **96**, 4834–4842 (1974).

12

SUBSTRUCTURE SEARCHING

12.1 PRINCIPLES OF SUBSTRUCTURE SEARCHING

Chemical information systems are the repositories for chemical structures and associated information that has been collected and archived. Once an effective method has been developed for storing chemical structures in a computer system, and once the method has been used to generate a file of chemical structures, the next step is to consider the uses for such a file.

One reason for the existence of a chemical information system is the ability to retrieve useful information from it. Among the reasons for a chemist to go to a chemical information system to ask questions are the following:

Has my query compound ever been reported?

Show me the compounds in the file that are most similar to my query compound.

What properties are known for my query compound?

How can my query compound be synthesized?

What compounds have similar substructural features as my query compound?

Many of these questions are essentially structural, that is, the heart of the question is a structure or a partial structure. The ability to carry out chemical substructure searching is an important attribute of chemical information systems.

The essence of substructure searching is simple, the identification of all the compounds in a given file that contain the specified substructural feature. Substructure searching is a special case of the more general problem of subgraph isomorphism, which involves the determination of whether one graph is a subgraph of another graph. The effectiveness of procedures to accomplish substructure searching depends on many factors, such as the following:

1. The method used for representation of the structures in the file being searched.
2. The method used for representation of the query substructural features.
3. The sophistication of the algorithm employed.
4. The size of the structure file being searched.

We will investigate the effect of each of these factors in the following pages.

Two measures of value for substructure search systems are *precision* and *recall*. Recall is the fraction of the actual matches that were found. If a substructure search yields all of the structures that it should have found, recall is 100%. Precision is the fraction of the retrieved structures that are relevant. If a substructure search yields 50 structures but only 25 of them are relevant, precision is 50%.

Substructure searching of files of structures is a computationally complex and expensive task. Substructure searching is a specific example of a class of problems known as the subgraph isomorphism problem, which deals with the determination of whether or not a given graph is imbedded within another graph. The subgraph isomorphism problem is known to be NP complete, which means that the increase of execution time of any algorithm increases as a polynomial function of the graph size. In adition, the time required for ordinary inputs grows rapidly with the size of the problem. This leads to the necessity of breaking down substructural searching into two component tasks.

The individual components of the overall substructure searching task are shown in Figure 1. The structural entries that are potential matches for the query substructure must be sifted out of the main structure file and only then searched in complete detail. The initial sifting is done with *screens*. This simple and fast sifting method eliminates the vast majority of the structures from further consideration. The screens represent a set of structural characteristics that can be identified both in the query substructures and the structures being searched. The screen search of the structure file picks out all those structures that match the screens of the query substructure.

The cost and response time of an on-line substructure searching system are determined to a large extent by the following two factors:

1. The number of structures from the file being searched that are passed by the screen search.
2. The number of actual valid answers to the substructural query posed.

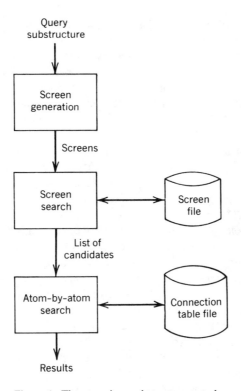

Figure 1. The steps in a substructure search.

Thus, the design and use of effective screens is extremely important to the overall effectiveness of a substructural search system.

Screen Generation

Screen searching is implemented by defining a set of structural features that characterize the structures in the file and the substructural query. The screens are derived from the connection table representations of the structures in the file just once, and they are stored in a screen file. The screens for the query substructure are derived as the first step in a substructure search. The objective in defining a set of screens is to encode as much structural information about the relevant structure or substructure as possible. However, the screen set must not be unduly long or computationally expensive to determine.

Simple screens can be derived directly from the connection table of a structure because the connection table contains information about the atoms, bonds, and connections present in the compound. However, simple screens are not sufficient to screen out enough compounds from a file during searching.

A much more effective set of screens can be derived by the use of chemically significant fragments as screens. Such fragments can also be generated from the connection tables by software. The set of chemically significant screens to be used in a particular system can be decided by algorithm. The molecules of the file are fragmented by a given set of rules. Thus, the fragments generated reflect the characteristics of the structures in the file, and both the number and type of fragments generated are characteristic of the file.

An example of a set of easily determined information that might constitute a screen set is shown in Figure 2. Atom type, bond type, and ring counts are simple screens that characterize a molecular structure to some degree. Thus, if a substructure containing two rings, or bromine, or nitrogen were being matched against a hydrocarbon containing only carbon and hydrogen, a simple check of the screens would eliminate it from further consideration. Screens can also note the presence, or absence, of small simple substructures such as those shown in Figure 2. If a potential candidate structure lacks one of the necessary screens of the query substructure, it can be eliminated from further consideration. Obviously, the intent in screening is to achieve 100% recall with 100% precision. If this were feasible, there would be no need for the further refinement of the searching procedure, the atom-by-atom iterative search.

If the individual pieces of structural information can be reduced to bits, an entire screen set can be stored in a binary string. This leads to very great efficiency since the screen set for a candidate structure and the screen set for the query substructure can then be compared using logical functions. This is, in fact, the common practice in actual searching systems.

For files of compounds of even moderate size, the total number of possible screens is very large. Accordingly, some criteria must be adopted for deciding which screens to use. One important criterion is that, if possible, each screen should split the file of compounds in half. That is, half of the compounds in the file should have a value for the screen of 0 and half

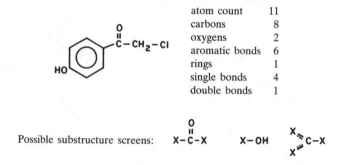

atom count	11
carbons	8
oxygens	2
aromatic bonds	6
rings	1
single bonds	4
double bonds	1

Figure 2. Example structure and example screens.

should have the value of 1. Information theory tells us that this would allow the maximum amount of information to be stored in the set of screens.

A great deal of information can be coded in binary form. Screens can be designed to code nearly any structural information in this form. As examples, consider the following potential screens:

Screen for carbon number:
   0 if carbon number is less than the mean
   1 of carbon number is greater than the mean

Screen for nitrogen:
   0 if no N present
   1 if at least one N present

Screen for oxygen:
   0 if no O present
   1 if at least one O present

Screen for molecular size:
   0 if the molecular weight is less than the mean
   1 if the molecular weight is greater than the mean

Screen for a substructure, say carbonyl:
   0 if carbonyl moiety absent
   1 if present

Screen for aromaticity:
   0 if no aromatic bonds present
   1 if aromatic bonds are present

Screen for five-membered rings:
   0 if no five-members rings are present
   1 if at least one five-membered ring is present

Screen for benzene ring:
   0 if no six-membered aromatic rings present
   1 if at least one six-membered aromatic ring present

Note that the definitions of some of the screens are derived from the properties of the structural file that is to be searched. Thus, the screens are custom designed to sort as efficiently as possible through the file. In the screen examples above, the carbon number example and the molecular weight example show this property of tailoring.

Screen generation algorithms much more elaborate and efficient to use have been devised. For example, Willett (1979) has described a screen set generation method that develops screens of equal frequency that are structural and are very efficient to employ. The Chemical Abstracts Service, which must deal with a file of structures in the millions, has devised an extremely elaborate screen set with more than 6000 different screen definitions (Dittmar et al. 1983).

The CAS ONLINE search system has a dictionary of screens that are used by the system. The screens have good selectivity as well as chemical significance. There are 12 types of screens employed in three classes: (1) augmented atom screens that describe atoms and their nearest neighbors and include augmented atom (AA), hydrogen augmented atom (HA), and twin augmented atom (TW) screens; (2) linear sequence screens that describe linear strings of atoms (excluding hydrogens) and are atom sequence (AS), bond sequence (BS), and connectivity sequence (CS) screens; (3) general structural feature screens that include ring count (RC), type of ring (TR), atom count (AC), degree of connectivity (DC), element composition (EC), and graph modifier (GM) screens. Augmented atom screens are descriptions of atoms and their nearest neighbor attachments (excluding hydrogens), for example. Atom sequence screens are descriptions of linear sequences of four, five, or six atoms (excluding hydrogen). A screen set of this complexity and sophistication is necessary for CAS ONLINE because of the size of the file being searched.

Atom-by-Atom Search

The matching of a substructure against a structure by an atom-by-atom search is, as mentioned above, a subgraph isomorphism problem. It is simple in conception but complicated in practice. The method begins with the least commonly occurring atom in the substructure and looks through the structure for an equivalent atom. When an equivalent atom is found, the next nearest neighbor atoms of the starting atom are checked for equivalence. When equivalences are found, the next nearest atoms are checked. Whenever an equivalence is not found, the algorithm backs up to the last successful checking point. The algorithm exhaustively tests all such correspondences between the atoms of the substructure query and the structure until either (a) a match is found or (b) all possible correspondences have been checked.

Consider a specific example of an atom-by-atom search as shown in Figure 3. The structure shown is to be examined in order to determine whether the substructure is present. Say that the search starts with the carbonyl carbon of the substructure, atom 1 in the substructure. The atoms of the structure would be searched, starting with atom 1, until a carbon atom is found, here atom 1 in the structure. Then the next nearest neighbors of substructure atom 1 would be compared to the next nearest neighbor atoms of atom 1 in the structure. Atom 2 in the substructure is equivalent to atom 4 in the structure, and the bond between atoms 1 and 2 in the substructure is double, as is the bond between atoms 1 and 4 in the structure. Atom 3 of the substructure is then compared to atom 2 in the structure, and they are found to be different. The algorithm backtracks to the last correct equivalence, and it next checks atoms of the substructure to atom 3 of the structure. Equivalence is found. Finally, atom 4 of the substructure is compared to atom 6 of the structure, and they are found to

Figure 3. Example structure and substructure being sought by atom-by-atom search.

be equivalent. All the atoms of the substructure have been matched against atoms in the structure, and in the proper bonding configuration, so the substructure has been found in its entirety. The search of this structure is complete, and the algorithm can now go on to the next structure in the file.

The amount of backtracking done in an atom-by-atom search depends on the amount of symmetry in the structure, the nature of the substructure, the numbering of each, and many other factors. The overall speed of the search depends on the amount of backtracking as well as the sheer size of the substructure. The potential complexity of the atom-by-atom search makes it evident why screening is necessary.

Query Input and Output of Results

Chemical information systems that support substructure searching also support graphical input and output. Graphical description of molecular structures and substructures is the common language of chemistry, so it follows that users of chemical information systems would want to express their queries graphically and receive their answers graphically. Essentially all working chemical information systems that support substructure searching also support graphical input and display. These two tasks are completely dependent on the hardware available as well as on the software being used.

The purpose of graphical input of substructures is to let the chemist user express a query in its natural language. The software can translate the query into the connection table storage required to proceed with the search. Graphical input systems have been built with light pens, cursor control, mouses, and graphics tables. In each case, the software is designed to mimic the motions of drawing a structure on a blackboard. There is, however, at least one substantial advantage in computer software drawing—previous structures can be recalled to the screen for modification. Each new structure or substructure need not be drawn from scratch.

The output of the results of searching can be as simple or sophisticated as the software designer wishes and as the hardware available allows. The usual output is the familiar sticklike structural drawing, once again, similar to blackboard drawings. A discussion of the types of graphical displays available for presentation of chemical structures will be given in Part IV.

12.2 A WORKING SYSTEM: CAS ONLINE

The Chemical Abstracts Service, an arm of the American Chemical Society, has been collecting abstracts of the chemical literature since before the turn of the century. The process has been computerized since 1965, with all the stored information accessible to computer search through CAS ONLINE more recently. The Chemical Registry System files contain more than six million substances.

One popular type of search query involves the definition of the search query in terms of a structural diagram. The diagram can be defined by the user on a graphics terminal or an alphanumeric terminal. The exact way in which the query is input is dependent on the hardware available to the user. The CAS ONLINE system can interact with many of the popular graphical and alphanumeric terminals. A query to the system can consist of one structural diagram or several structural diagrams. The Boolean logical operators AND, OR, and NOT can be used in conjunction with the query parts. Screens can also be constructed by the user; these are features that must be present in the sought structures in addition to the structural diagrams.

The CAS ONLINE system sifts the file being searched with the screens supplied by the user in addition to its own screens. It then applies an iterative atom-by-atom search to those structures passing the screening process. The user can request the CAS registry numbers, full structural diagrams, and other information about any of the substances that are retrieved by the search.

REFERENCES

J. E. Ash, P. A. Chubb, S. E. Ward, S. M. Welford, and P. Willett, *Communication, Storage, and Retrieval of Chemical Information*, Wiley, New York, 1985.

P. G. Dittmar, N. A. Farmer, W. Fisanick, R. C. Haines, and J. Mockus, "The CAS ONLINE Search System. 1. General System Design and Selection, Generation, and Use of Search Screens," *J. Chem. Inf. Comput. Sci.*, **23**, 93–102 (1983).

M. F. Lynch, J. M. Harrison, and W. G. Town, *Computer Handling of Chemical Structure Information*, MacDonald, London, 1971.

M. F. Lynch, J. M. Barnard, and S. W. Welford, "Generic Structure Storage and Retrieval," *J. Chem. Inf. Comput. Sci.*, **25**, 264–270 (1985).

R. E. Stobaugh, "Chemical Substructure Searching," *J. Chem. Inf. Comput. Sci.*, **25**, 271–275 (1985).

R. E. Tarjan, "Graph Algorithms in Chemical Computation," *A.C.S. Symp. Ser.*, **46**, 1–20 (1977).

P. Willett, "A Screen Set Generation Algorithm," *J. Chem. Inf. Comput. Sci.*, **19**, 159–162 (1979).

W. T. Wipke, S. R. Heller, R. J. Feldmann, and E. Hyde, *Computer Representation and Manipulation of Chemical Information*, Wiley, New York, 1974.

13

MOLECULAR MECHANICS

13.1 INTRODUCTION

Once the topology of a chemical structure is specified, the next level of information is the conformational or geometric properties of the structure. Chemists commonly use mechanical models of chemical structures as an aid to visualization. Looking at molecules as three-dimensional objects naturally leads to consideration of strain energies, that is, the relationship between conformation and the energetic requirements to adopt a particular geometry. For the past 40 years, chemists have relied largely on mechanical models to aid them, but these models have limitations that can have undesirable effects on one's reasoning. Fortunately, methods for calculating molecular geometries and associated energies are now available and can be used as a routine research and teaching tool. The method of molecular mechanics was introduced by Westheimer (1956) and refined by Wiberg (1965). It is known as the strain energy minimization technique, the molecular mechanics method, or the force field method. This chapter describes this important area of modern chemical research.

Basis of Method

The fundamental assumption of molecular mechanics is that data determined experimentally for small molecules (bond lengths, bond angles, etc.) can be extrapolated to larger molecules. A molecule is considered to be a collection of atoms held together by simply elastic or harmonic forces. The forces are defined in terms of potential energy functions related to the internal coordinates of the molecule. The forces collectively make up what is called the force field of the molecule. A molecule at rest in its force field will adopt the lowest energy conformation that is accessible. The energy of any conformation can be calculated from a knowledge of the force field and of the coordinates of the atomic constituents of the molecule. The energy thus calculated is often called the strain energy:

$$E_{strain} = E_{bond} + E_{angle} + E_{torsional} + E_{nonbonded} \tag{1}$$

It is the sum of contributions from terms arising from bond length stretching or compressing, bond angle bending, torsional angle twisting, nonbonded interactions, and possibly others. Each interaction (e.g., bond length) has an optimum vlaue. Each term in the summation can be parameterized in many different ways with different functional forms and parameter values, and many chemists have published their ideas in this area. Force fields have been developed in order to reproduce structural geometry, relative conformational energies, heats of formation, crystal packing arrangements, and other properties. Each term in the equation is actually a summation over all occurrences of that particular interation in the molecule.

In practice, the strain energy of a molecule is minimized by a numerical method such as gradient search whereby the atomic coordinates are altered and the energy recalculated repeatedly. The objective of the computation is to find the conformation with the least strain energy spread throughout the structure. Force field methods provide a rigid structure at rest in an energy minimum.

The broadest objective of molecular mechanics is to obtain a simple but widely applicable force field model backed up with experimental observations that can accurately generate the structures (and thermodynamic properties) of organic molecules. This objective has been fulfilled to a significant degree, but work continues in a search for further improvements in accuracy and generality.

13.2 IMPLEMENTATION OF THE METHOD

The Force Field

A variety of functional forms and parameters have been used to construct force fields. Perhaps the most widely used force field has been developed by Allinger and coworkers (Burkert and Allinger 1982). This force field is embodied in software available from the Quantum Chemistry Program Exchange.

Each type of intramolecular interaction included in the strain energy equation is represented by a functional form that includes a restoring force. Each term is formulated so that it tends to decrease in magnitude as the bond length or angle or other measure tends towards its optimum value.

The terms entering into the strain energy calculation can be put into three classes:

1. Forces that involve two atoms such as bond length deformation and nonbonded interactions.

2. Forces that involve three atoms such as bond angle deformations.

3. Forces that involve four atoms such as torsional angle deformations.

Bond Length. The atoms that are joined by each bond are considered to be masses joined by a spring. Hooke's law is used to represent the energy necessary to stretch or compress a bond from its optimum length:

$$E_{bond} = \sum_{i=1}^{N_{bond}} k_i(l_i - l_i^0)^2 \tag{2}$$

where N_{bond} is the number of bonds in the molecule, k_{ij} is a proportionality constant (dependent on the bond type and atom identities), l_i is the actual bond length of bond i, and l_i^0 is the optimum bond length (dependent on the bond type and atom identities). The total bond length deformation energy contribution to the strain energy is taken to be the sum over all the bonds in the molecule.

Bond Angles. A Hooke's law formulation is also used for the representation of bond angle strain:

$$E_{angle} = \sum_{i<j}^{N_{angle}} k_{ij}(\theta_{ij} - \theta_{ij}^0)^2 \tag{3}$$

where N_{angle} is the number of bond angles present in the molecule, k_{ij} is a proportionality constant (dependent on the bond types and identities of the three atoms involved), θ_{ij} is the actual bond angle involving bonds i and j, and θ_{ij}^0 is the optimum bond angle (dependent on the bond types and identities of the three atoms involved). The total contribution to the strain energy by bond angle deformation is taken as the sum over all the angles in the molecule.

The amount of energy necessary to bend a bond angle is less than that required to stretch a bond, so the force constant in Equation (2) is larger (by approximately an order of magnitude) than that of Equation (3). Therefore, a structure is more likely to have strained angles than strained bond lengths when it is in a conformation of low total strain.

Hooke's law is quadratic, so the contributions to strain energy from an equal-increment deformation are larger as the bond angle moves further from the optimum value. A better fit between strain and deformation would be possible if the force constant were made a function of $\Delta\theta = |\theta_{ij} - \theta_{ij}^0|$. An alternative way to correct the terms, which has been employed in some force fields, is to include a cubic term in Equations (2) and (3).

Including independent terms for bond length and bond angle deformations neglects the obvious interplay between these aspects of molecular conformation. As a bond angle is changed, the two associated bond lengths will tend to change also. This source of strain will appear in the calculation of energies between atoms in a 1–3 relationship (atoms each bonded to a common atom) due to through-space interactions if such a term is included in the force field. Alternatively, a cross term that includes both bond length and bond angle deformations multiplied together can be used.

Torsional Angles. The torsional angle term represents the energy cost of rotation about single bonds. The interaction involves three bonds and four atoms. Call the atoms A, B, C, and D (Figure 1). The torsional angle is defined as the angle measured about the B–C axis from the A–B–C plane to the B–C–D plane. There are two possible conventions as to which direction of rotation is defined as a positive angle and which is negative, and both conventions have been used. As long as a particular force field is consistent, the convention employed is immaterial.

The functional form most widely used to represent the torsional strain has been a truncated Fourier series:

$$E_{torsion} = \sum V_1(1 + \cos \omega) + V_2(1 - \cos 2\omega) + V_3(1 + \cos 3\omega) \qquad (4)$$

This funtion has local maxima at 0° and 120° and local minima at 60° and 180°, which mimics the observed torsional barrier of alkanes. The summation is taken over all possible torsional angles in the molecule. The constants V_1, V_2, and V_3 are chosen so that the force field reproduces known conformations for simple molecules. In general, it takes less energy to deform torsional angles than either bond lengths or bond angles.

Nonbonded Interactions. As the internuclear distance between two atoms varies, a potential energy (either attractive or repulsive) is generated. As the

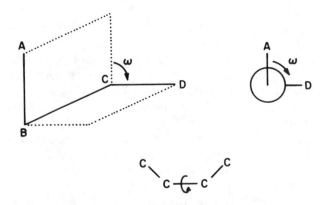

Figure 1. Torsional angle conventions.

atoms approach one another, an attraction due to London dispersion forces acts on them. As they come closer, a van der Waals repulsion occurs (see Figure 2).

Two functions that describe this phenomenon are the Lennard-Jones 12–6 potential and the Buckingham exponential–6 potential:

$$V_{LJ} = \epsilon \left[\left(\frac{R_0}{R} \right)^{12} - 2 \left(\frac{R_0}{R} \right)^{6} \right] \tag{5}$$

$$V_{B} = \frac{\epsilon}{1 - 6/\alpha} \left[\frac{6}{\alpha} \exp \left[\alpha \left(1 - \frac{R}{R_0} \right) \right] - \left(\frac{R_0}{R} \right)^{6} \right] \tag{6}$$

where ϵ is the depth of the potential well, R is the distance between atoms at which the potential function is minimized, and α is related to the steepness of the repulsive part of the interaction. When α has a value of 14–15, the steepness of the repulsive term in the Buckingham potential is similar to that of the Lennard-Jones 12–6 potential. Both of these functions describe the attractive interactions with inverse-6-dependent terms, but they describe the repulsion interactions differently. The Lennard-Jones potential function has two adjustable parameters, and the Buckingham potential has three.

A potential is calculated for all pairs of atoms in the molecule, except those bonded together or those bonded to a common third atom.

The values for the parameters incorporated into nonbonded interaction potentials are derived from experimental observations in crystals. These parameters are assumed to be applicable to the intramolecular interactions being dealt with in molecular mechanics. Another implicit assumption is that the potentials are pairwise additive. The induced dipole moments of the two atoms can be altered by any other nearby atoms. However, the potential developed by two atoms is usually considered to be independent of what other atoms might be nearby (even if a third atom is interspersed between the two atoms of interest). The spherical symmetry of these nonbonded interaction functions is also a drastic simplification of reality. Although these assumptions can be faulted, the necessity to keep the force field relatively simple for computational reasons has led to their adoption.

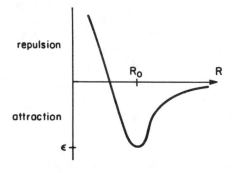

Figure 2. Shape of nonbonded potentials.

Minimization Methods

The basic task in the computational portion of molecular mechanics is to minimize the strain energy of the molecule by altering the atomic positions. This means minimizing the nonlinear strain energy represented by Equation (1) with respect to the independent variables, which are the Cartesian coordinates of the atoms. A number of different mathematical methods have been utilized for this purpose in molecular mechanics. Altona and Faber (1974) discuss five methods: steepest descent, parallel tangents, pattern search, nonsimultaneous local energy minimization, and valence force minimization. The most widely used methods fall into two general categories, steepest descent and related methods, which use first derivatives, and Newton–Raphson procedures, which additionally use second derivatives.

Steepest descent was the first method used by Wiberg (1965) and studied extensively by Allinger and coworkers (Burkert and Allinger 1982). This method depends on (1) either calculating or estimating the first derivative of the strain energy with respect to each coordinate of each atom and (2) moving the atoms. The derivative is estimated for each coordinate of each atom by incrementally moving the atom and storing the resultant strain energy change. The atom is then returned to its original position, and the same calculation is repeated for the next atom. After all the atoms have been tested, their positions are all changed by a distance proportional to the derivative calculated in step 1. The entire cycle is then repeated. The calculation is terminated when the energy is reduced to an acceptable level. The main problem with the steepest descent methods is that of determining the appropriate step size for atom movement during the derivative estimation steps and the atom movement steps. The sizes of these increments determines the efficiency of minimization and the quality of the result.

The methods known as parallel tangents and pattern search are actually variants of the steepest descent method. The parallel tangents procedure starts by finding two new locations for an atom each of which has lower strain energy than the initial point. Then a curve is fit through the three points, its minimum is found, and this fourth point is chosen for the new atom location. The cycle repeats. The pattern search method is like the steepest descent method except that each atom is left at its best position as positions are tested. An additional advancement involves altering the step sizes taken by the atoms as dictated by the size of the derivative found during the search or by the estimated value of the second derivative.

An advantage of these first-derivative methods is the relative ease with which the force field can be changed. If the first derivatives are being estimated numerically, only the potential function itself appears in the programs. Therefore, it can be altered without ramifications elsewhere in the program.

The Newton–Raphson methods of energy minimization utilize the curvature of the strain energy surface to locate minima. The computations are considerably more complex than the first-derivative methods, but they utilize the available information more fully and therefore converge more quickly. These methods involve setting up a system of simultaneous equations of size $(3n - 6) \times (3n - 6)$ and solving for the atomic positions that are the solution of the system. Large matrices must be inverted as part of this approach. The Newton–Raphson methods are discussed in detail by Burkert and Allinger (1982).

Graphical Display of Molecules

An important capability of computers is their ability to present complex information in pictorial form. In the context of molecular mechanics, this translates into the ability to display molecules graphically on video display terminals. Presentation of molecules for visual analysis allows the chemist to view the model and judge its quality or seek insights based on the structure. The topic of graphical display of molecular structures is covered in detail in Part IV.

REFERENCES

N. L. Allinger, "MM1: Molecular Mechanics Program," Quantum Chemistry Program Exchange, Program Number 318.

C. Altona and D. H. Faber, "Empirical Force Field Calculations. A Tool in Structural Organic Chemistry," *Top. Curr. Chem.*, **45**, 1–38 (1974).

D. B. Boyd and K. B. Lipkowitz, "Molecular Mechanics, The Method and Its Underlying Philosophy," *J. Chem. Ed.*, **59**, 269–274 (1982).

U. Burkert and N. L. Allinger, *Molecular Mechanics*, American Chemical Society, Washington, DC, 1982.

P. Kollman, "Theory of Complex Molecular Interactions: Computer Graphics, Distance Geometry, Molecular Mechanics, and Quantum Mechanics," *Acc. Chem. Res.*, **18**, 105–111 (1985).

B. Venkataraghavan and R. J. Feldmann (Eds.), *Macromolecular Structure and Specificity: Computer-Assisted Modeling and Applications*, New York Academy of Sciences, New York, 1985.

F. H. Westheimer, in *Steric Effects in Organic Chemistry*, M. S. Newman (Ed.), Wiley, New York, 1956, Chapter 12.

K. B. Wiberg, "A Scheme for Strain Energy Minimization. Application to the Cyclohexanes," *J. Amer. Chem. Soc.* **87**, 1070–1078 (1965).

14

PATTERN RECOGNITION

14.1 PATTERN RECOGNITION METHODS

Chemists are accustomed to using graphical methods for data interpretation. Typically, a set of measurements of one dependent variable is made for different values of one independent variable while the remaining independent variables are held constant. The results are plotted for viewing and analysis. If the variables are related through a simple relationship, then a curve (perhaps even a straight line) can be fit to the set of data. This operation constitutes a generalization and an interpretation. The set of data is characterized tersely by the fit equation's coefficients. The equation can also be used for prediction. Of course, the predictions of the dependent variable are most likely to be reasonable if the equation is used for interpolation between observed values of the independent variables. However extrapolation may also be warranted in some circumstances.

Situations arise in science, however, in which independent variables cannot be varied one at a time. There may be a very large number of independent variables to sort through. The data may be characterized by membership in categories rather than have a quantitatively measurable property. For example, a sample of stone comes from quarry 1 or quarry 2; it can never come from quarry 1.37. In many cases, this type of data can still be represented as points, but no longer in a two-dimensial space. An object or observation characterized by n descriptors can be represented as a point in an n-dimensional space.

Some examples of data that can be expressed in this notation are as follows:

1. Air pollution particulate samples characterized by the trace metal concentrations.
2. Trace-level organic acid concentrations in human body fluids.

Thus, the data encountered in chemical problems can often be represented by a vector as

$$\mathbf{X}_i = (x_1, x_2, \ldots, x_n) \tag{1}$$

where each individual component of the pattern vector, x_j, is the value for the jth descriptor. To reference the ith observation of the data set, the notation $\mathbf{X}_i$ is used. An equivalent way to view this representation of data is to consider each observation to be represented by a point in n-dimensional space.

For a given observation, which is represented by a given point, the value of each coordinate is just the numerical value for one of the descriptors comprising the representation. The expectation is that the points representing objects of a common category will cluster in one limited region of the space separate from those belonging to a different category. The clusters are regions of high local density relatively far apart from each other. *Pattern recognition* consists of a set of methods for investigating data represented in this manner to assess the degree of clustering and general structure of the data space.

Two classes of pattern recognition methods are supervised and unsupervised. Supervised methods employ a training set of patterns belonging to known classes to develop discriminants. These discriminants are then available for predicting the classes of unknown patterns. Unsupervised methods do not have access to the class information but only to the patterns themselves. The goal is to investigate the pattern space by looking for clusters of pattern points that may be indications of meaningful relations among the clustered points.

In order to discuss clustering, one must have a definition of similarity. It is most common to use Euclidean distance in the definition of similarity in pattern recognition work. The Euclidean distance between the two patterns

$$\mathbf{X}_i = (x_{i1}, x_{i2}, \ldots, x_{in}) \tag{2}$$

and

$$\mathbf{X}_j = (x_{j1}, x_{j2}, \ldots, x_{jn}) \tag{3}$$

is given by the expression

$$d_{ij} = \sum_{k=1}^{n} (x_{ik} - x_{jk})^2 \tag{4}$$

The similarity between the two patterns is inversely related to the distance between them. Different inverse relations have been used in different types of pattern recognition work, but a very common form is

$$S_{ij} = 1 - \frac{d_{ij}}{d_{max}} \qquad (5)$$

where d_{max} is a normalizing factor. If d_{max} is the largest interpoint distance in the data set, S_{ij} spans the range from unity for identical patterns to zero for the least similar pair of patterns.

Example Data Set

The most effective way to introduce many of the concepts of pattern recognition will be to use a specific data set for illustration. Therefore, a set of 100 patterns each consisting of five descriptors is presented in Figure 1. The first column refers to the class membership of each pattern, with 1 denoting the positive class and -1 denoting the negative class. There are 50 patterns in each class. The values of the raw descriptors range from 1 to 20.

1	9. 446	4. 652	5. 089	6. 715	7. 659
1	2. 275	7. 488	5. 846	3. 161	6. 999
1	8. 429	8. 758	6. 035	4. 489	3. 155
1	4. 399	6. 073	7. 395	1. 627	2. 080
1	6. 291	8. 199	1. 430	1. 651	5. 536
1	6. 674	1. 946	5. 121	6. 372	1. 194
1	1. 749	4. 944	8. 628	2. 311	5. 513
1	3. 048	6. 231	5. 220	9. 610	3. 046
1	2. 908	4. 805	3. 870	5. 941	5. 363
1	6. 922	4. 930	1. 826	2. 480	9. 829
1	7. 476	3. 589	9. 231	2. 259	5. 733
1	5. 756	9. 958	7. 919	7. 227	1. 535
1	4. 175	2. 696	3. 936	3. 557	5. 674
1	2. 871	9. 425	4. 766	7. 312	9. 847
1	8. 577	5. 058	1. 047	7. 018	3. 870
1	4. 348	6. 583	5. 728	2. 982	2. 992
1	6. 689	8. 739	4. 492	8. 297	3. 041
1	9. 455	7. 800	2. 158	2. 379	9. 298
1	9. 266	9. 077	2. 034	7. 627	6. 522
1	5. 503	4. 550	6. 226	4. 038	8. 867
1	7. 345	1. 293	3. 187	3. 292	2. 449
1	2. 352	8. 532	1. 901	8. 235	8. 630
1	3. 109	7. 563	3. 125	2. 462	8. 009
1	4. 075	9. 100	4. 393	7. 727	5. 511
1	2. 555	4. 735	5. 885	5. 698	7. 299
1	6. 464	3. 300	6. 113	6. 721	3. 521
1	1. 567	9. 808	5. 330	9. 532	3. 487
1	8. 783	5. 136	6. 777	1. 782	9. 780
1	6. 609	4. 111	4. 837	9. 836	3. 968
1	6. 879	5. 609	3. 993	3. 646	3. 119
1	9. 439	7. 733	6. 781	4. 484	9. 830
1	1. 654	7. 676	7. 608	3. 614	2. 206
1	3. 734	5. 677	5. 274	7. 038	3. 660
1	2. 135	9. 473	8. 350	9. 809	5. 401
1	6. 344	3. 960	4. 158	3. 321	5. 287
1	2. 006	8. 591	9. 868	7. 118	9. 357
1	1. 592	9. 357	9. 688	3. 748	8. 289
1	3. 594	9. 431	9. 278	8. 999	1. 018
1	9. 445	4. 268	3. 223	8. 712	7. 521

Figure 1. One hundred patterns of five descriptors each.

1	2. 260	4. 152	6. 866	1. 817	9. 250
1	7. 725	6. 020	4. 718	8. 456	1. 520
1	2. 125	2. 623	2. 794	1. 176	2. 070
1	7. 270	8. 248	1. 466	5. 155	6. 818
1	2. 828	1. 552	7. 244	5. 567	4. 739
1	5. 712	8. 347	3. 220	9. 281	1. 201
1	5. 004	9. 805	4. 983	9. 737	8. 059
1	9. 608	3. 267	7. 283	9. 118	6. 363
1	2. 467	5. 562	4. 090	3. 670	2. 879
1	6. 401	1. 149	8. 228	1. 455	4. 615
1	4. 134	2. 049	8. 256	2. 448	7. 241
−1	1. 047	7. 768	1. 232	9. 566	9. 657
−1	1. 008	5. 689	2. 309	9. 151	5. 504
−1	8. 184	2. 320	1. 038	9. 559	3. 555
−1	5. 382	1. 254	7. 743	9. 787	3. 955
−1	2. 261	4. 320	4. 996	7. 056	9. 340
−1	3. 579	2. 974	1. 185	7. 190	6. 958
−1	4. 085	1. 623	3. 737	7. 432	5. 965
−1	2. 146	4. 247	1. 190	4. 719	9. 541
−1	1. 177	1. 362	1. 915	8. 731	6. 713
−1	1. 157	4. 540	4. 673	7. 268	6. 750
−1	2. 749	1. 998	1. 341	5. 190	8. 152
−1	8. 958	2. 086	2. 452	9. 245	5. 935
−1	2. 093	1. 533	9. 708	8. 097	7. 790
−1	3. 330	3. 770	1. 455	9. 212	8. 542
−1	2. 654	3. 293	7. 049	9. 879	3. 001
−1	3. 726	2. 921	4. 516	9. 968	4. 082
−1	3. 092	1. 436	9. 442	9. 286	4. 120
−1	4. 378	1. 066	2. 362	7. 397	5. 232
−1	8. 009	2. 085	1. 943	9. 197	4. 167
−1	1. 101	3. 529	1. 657	5. 419	4. 061
−1	7. 158	1. 457	1. 034	9. 176	9. 725
−1	1. 317	1. 700	2. 500	9. 372	5. 807
−1	2. 056	4. 475	1. 524	6. 913	7. 857
−1	3. 025	1. 310	6. 264	6. 598	5. 429
−1	1. 737	8. 103	2. 569	9. 755	9. 521
−1	3. 501	2. 109	3. 246	9. 058	3. 312
−1	2. 221	4. 699	3. 316	7. 243	6. 017
−1	2. 279	1. 586	9. 509	9. 885	2. 946
−1	2. 623	1. 391	4. 690	8. 046	8. 887
−1	5. 128	2. 969	1. 016	9. 746	8. 041
−1	2. 153	2. 851	8. 124	9. 884	5. 044
−1	4. 951	3. 286	4. 621	9. 574	3. 964
−1	3. 533	1. 038	9. 379	8. 235	5. 738
−1	1. 274	2. 449	3. 598	7. 628	2. 649
−1	2. 007	3. 298	2. 301	9. 787	1. 103
−1	5. 083	1. 162	3. 390	8. 351	9. 660
−1	2. 302	2. 236	1. 904	8. 819	3. 910
−1	2. 200	1. 134	2. 263	3. 299	7. 722
−1	1. 737	1. 964	2. 830	5. 420	8. 458
−1	1. 202	4. 833	2. 748	9. 669	6. 601
−1	3. 478	3. 282	1. 898	5. 178	8. 770
−1	6. 294	1. 508	4. 888	9. 868	1. 885
−1	1. 394	1. 421	2. 657	6. 356	6. 314
−1	2. 297	1. 351	3. 578	8. 646	4. 493
−1	4. 130	2. 838	8. 205	9. 652	6. 082
−1	5. 227	3. 246	1. 928	7. 250	4. 956
−1	4. 458	1. 216	6. 749	8. 130	5. 664
−1	4. 399	2. 167	2. 067	5. 227	9. 735
−1	4. 763	1. 766	5. 146	9. 342	9. 912
−1	1. 562	3. 538	3. 987	6. 758	4. 492

Figure 1. (*Continued*).

Statistics relating to the descriptors are found in Table 1, part A. These statistics refer to the overall set of 100 patterns, without regard to class membership. The simple correlations between each pair of descriptors are shown in Table 1, part B. The small values that appear in the correlation matrix show that the descriptors are almost completely orthogonal to one another, that is, these five descriptors are nearly independent. Figure 2 depicts this visually. It is a plot of descriptor 1 versus descriptor 2, and it shows a scattered pattern of values.

These 100 patterns belong to two classes. The means and standard deviations for each class are shown in Table 1, part A. An interesting test is to see how well the patterns can be put into their proper categories based on one descriptor at a time. The results of such a test are shown in Table 1, part C. Thresholds were inserted into the descriptors to maximize the

TABLE 1. Statistics for the Five-Dimensional Data Set

A. Individual Descriptor Statistics

	Overall		Yes Class		No Class	
Descriptor	Mean	Standard Deviation	Mean	Standard Deviation	Mean	Standard Deviation
1	4.27	2.50	5.23	2.62	3.31	1.97
2	4.40	2.73	6.07	2.61	2.72	1.60
3	4.57	2.56	5.34	2.35	3.80	2.56
4	6.77	2.67	5.41	2.79	8.12	1.69
5	5.79	2.57	5.42	2.74	6.15	2.36

B. Correlation Matrix

	1	2	3	4	5
1	1.0	0.067	0.057	0.108	0.048
2		1.0	0.087	0.108	0.022
3			1.0	0.082	0.182
4				1.0	0.136
5					1.0

C. Classification by Individual Descriptors

	Number Incorrect		
Overall	Yes	No	Percentage
30	25	5	70
32	12	20	68
42	26	16	58
38	25	13	62
45	16	29	55

TABLE 1. (*Continued*)

D. Results of Eigenanalysis

	Eigenvalue	Cumulative Variance (%)
1	1.208	24.4
2	1.182	48.3
3	1.002	68.5
4	0.883	86.4
5	0.675	100.0

Eigenvector 1

$(-0.40004, -0.53842, -0.14597, 0.67739, -0.26438)$

Eigenvector 2

$(0.00252, 0.21052, 0.70373, 0.05655, -0.67620)$

E. Transformation of Pattern 1 and 39

Pattern 1 Descriptor Raw Value	Pattern 1 Autoscaled Value	Pattern 39 Descriptor Raw Value	Pattern 39 Autoscaled Value
9.446	2.0715	9.445	2.0711
4.652	0.09289	4.268	-0.047656
5.089	0.20351	3.223	-0.52512
6.715	-0.020294	8.712	0.72794
7.659	0.72931	7.521	0.67559

Pattern 1

Product of autoscaled values with eigenvector $1 = -1.115$
Product of autoscaled values with eigenvector $2 = -0.326$

Pattern 39

Product of autoscaled values with eigenvector $1 = -0.41158$
Product of autoscaled values with eigenvector $2 = 0.79030$

---

number of correct classifications without regard to class. Based on descriptor 1, 70% correct (30 patterns classified incorrectly) was the best result obtained. Thus, there is no one descriptor that does a very good job of separating these 100 patterns into their two classes. This set of 100 patterns will be used in the following pages to illustrate some of the pattern recognition methods discussed.

The methods and techniques of pattern recognition can be subdivided into the following categories for simplification: preprocessing, mapping and display, classification by discriminants, clustering methods, and modeling methods. Each of these will be introduced in the following pages.

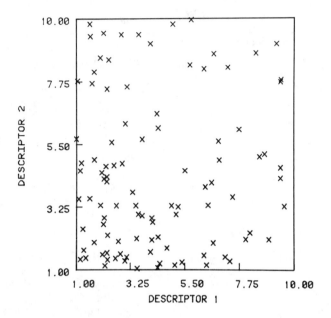

Figure 2. Plot of values for descriptor 1 versus descriptor 2 for the test data set.

Preprocessing

Preprocessing operations alter the form of the representations of the observations and include operations such as scaling, normalization, and other mathematical transformations. The objective of all preprocessing operations is to make the analysis phase of the pattern recognition study easier.

Scaling and normalization are required to convert the given units of measurements for different descriptors to a compatible form. For example, if one descriptor had a natural range of 1–10 and another had a natural range of 0.003–0.040, simultaneous analysis of these descriptors without scaling could give unwarranted and unwanted emphasis to one or the other of the descriptors. In addition, numerical instabilities in some algorithms can result from such imbalance among descriptor value ranges.

One preprocesssing method that is widely used to correct such imbalances is called autoscaling. This method simultaneously scales and normalizes the data by translation so that for each descriptor the mean becomes zero and the standard deviation becomes unity. The equation that describes auto-scaling is

$$x'_{ij} = \frac{x_{ij} - x_j}{\sigma_j} \tag{6}$$

where x_j is the mean for the jth descriptor over the data set, and σ_j is the standard deviation for the jth descriptor over the data set. The x_{ij} are the original, raw data points, and the x'_{ij} are the autoscaled data.

For descriptor 1 in the example data set, the mean value is 4.27 and the standard deviation is 2.50. Thus, the raw value of 9.45 for the first pattern would autoscale to $(9.45 - 4.27)/2.50 = 2.07$.

Mapping and Display

One powerful way of analyzing the structure of a data set is to map the points from the high-dimensional space to a two-dimensional space (a plane) for direct viewing. It has long been recognized that people are adept at recognizing patterns in two-dimensional plots, and this has been exploited in many pattern recognition display methodologies. A number of useful and imaginative graphical techniques exist for directly displaying multivariate data in just one or two dimensions. Their use has been neglected in the physical sciences in favor of more common rectangular plots and histograms. However, display techniques employing metroglyphs, linear and circular data profiles, Andrews plots, and Chernoff faces can be valuable tools for visualizing relationships between observations, for identifying outliers, and for classification purposes (e.g., Wang 1978).

A method that combines both mapping and display operations along with feature selection is principal components analysis, also known as Karhunen–Loéve transformations (Tou and Gonzalez 1974). This transformation involves performing an eigenanalysis on the variance–covariance matrix of the data set, and then the eigenvectors are used to perform a linear rotation. The multidimensional data set can be plotted in the two dimensions determined by the two principal eigenvectors and displayed for visual analysis. Alternatively, the data can be rotated by multiplication with any desired number of eigenvectors to reduce the dimensionality of the data set. This operation is commonly done using just enough of the available eigenvectors to retain 95 or 99% of the total variance of the data set. An advantage of this feature selection procedure is that the number of descriptors per observation is reduced, thus making further analysis more convenient in some ways. A disadvantage is that the individual descriptors resulting from this transformation are mixtures of the original descriptors, thus complicating analysis of later results.

When the example data set was submitted to a Karhunen–Loéve analysis, the results shown in Table 1, part D, were obtained. The eigenvalues were sorted by value from largest to smallest. The cumulative variance column shows that only 48% of the overall variance of this set of data is represented in the first two principal components. The almost equal size of the eigenvalues shows that this set of five-dimensional data nearly spans the space completely. The first two eigenvectors are shown. Each has been normalized to unit length. Vector multiplication of these two eigenvectors by the five-dimensional pattern vector for individual patterns of the example data set yields two-dimensional points, as shown in Table 1, part E. This is the Karhunen–Loéve transformation from 5-space to 2-space for individual patterns. When this operation is done for all 100 patterns, the results are as shown in Figure 3. This is called a principal components plot. The x axis of the plot represents principal component 1, which is the direction in the 5-space that contains the most variance. The relative percentage of the total variance explained by the first principal component is 24.4% as shown in Table 1, part D. The y axis of the plot, principal component 2, is that direction orthogonal to principal component 1 that contains the most additional variance of the data set, namely 23.9% Even though the data were not separable into their two classes to any appreciable degree based on any one descriptor, the principal components plot of Figure 3 shows a great deal of separation. Thus, the Karhunen–Loeve transform has mapped a five-dimensional space that could not be viewed directly to a two-dimensional one that could be viewed.

Another mapping strategy is to attempt to find a two-dimensional display where each point in the original data set is represented by a point in the two-dimensional plane. This involves attempting to represent a high-dimensional space directly in two dimensions. Of course, a certain amount of error in the mapping is to be expected. Many papers in the literature report approaches to the nonlinear mapping problem. Iterative nonlinear mapping routines have been implemented using the error function

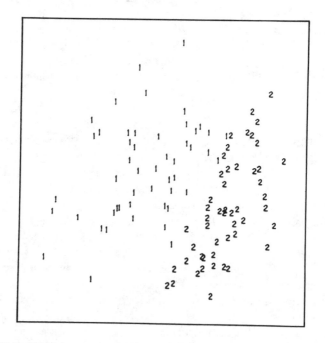

Figure 3. Plot of 100 test patterns in the space defined by their two principal components.

$$\frac{\Sigma \, (d_{ij} - d'_{ij})^2}{\Sigma \, d'_{ij}} \tag{7}$$

where d'_{ij} is the Euclidean distance between points i and j in the original n-dimensional space, and d_{ij} is the two-dimensional interpoint distance. The method of steepest descent, for example, can be used to minimize E. In a perfect mapping, where the n-space is perfectly represented in 2-space, E would be zero. Nonlinear mapping algorithms are notorious for consuming large amounts of computer time. The starting estimates of the 2-space point locations must be entered into the algorithm by the user. Both the final results obtained and the computer time invested in attaining the results are strong functions of the quality of the initial estimates entered. They might be the results of a previous principal components analysis, for example. The results of the nonlinear mapping algorithm are usually presented graphically for visual examination.

Classification by Discriminants

It is very common for data sets being studied by pattern recognition methods to be *category* data. That is, each observation is tagged by its membership in a discrete category. Then, pattern recognition methods are used to examine the points in the high-dimensional space to see if the point sets are disjoint. In other words, the goal is to discover if the points belonging to one category are in a cluster separated from the points belonging to another cluster. One way to ascertain if this structure is present involves putting discriminant surfaces through the space to slice it into regions. The search for powerful discriminants that indeed classify the points correctly comprises the operation of classification and discriminant development.

A discriminant function can also be thought of as a decision surface passing through the n-space that contains the points of the data set. If the surface is flat (that is, if it is a hyperplane), it can be completely and unambiguously represented by a normal vector perpendicular to it. With no loss of generality, the decision surface can be constructed to pass through the origin of the space. Then, every vector from the origin defines a plane that is the locus of points perpendicular to the normal vector. We will call the vector representation of a decision surface the *weight vector*.

Since it is very convenient to have the decision surface pass through the origin of the space, it is worthwhile to ensure that this will always be possible. An extra, orthogonal dimension is added to the pattern space, and all the pattern vectors are augmented with a $(d + 1)$st component. The $(d + 1)$st component of the pattern vectors can be assigned any value, but it is usually given the value of unity.

In addition to being able to represent a linear decision surface by specifying its normal vector, another important feature arises for this simple situation. The dot product of the normal vector **W** and a pattern vector **X** defines on which side of the hyperplane a given pattern point lies:

$$\mathbf{WX} = |\mathbf{W}||\mathbf{X}| \cos \theta \tag{8}$$

where θ is the angle between the two vectors.

$$\cos \theta > 0 \quad \text{for } -90° < \theta < 90°$$
$$\cos \theta < 0 \quad \text{for } 90° < \theta < 270° \tag{9}$$

Since the normal vector is perpendicular to the hyperplane decision surface, all patterns having dot products that are positive lie on the same side of the plane as the normal vector, and all those with negative dot products lie on the opposite side. Points with zero dot products lie in the plane, which constitutes another definition of the location of the plane. (In a computational sense, the probability of a point lying within the plane is very small, with the exact probability depending on the number of significant digits used in the floating-point system of the computer being employed.) A second and equivalent way to define the dot product of two vectors is

$$\mathbf{WX} = |\mathbf{W}||\mathbf{X}| \cos \theta = w_1 x_1 + w_2 x_2 + \cdots + w_d x_d + w_{d+1} \tag{10}$$

Each of the components of the weight vector **W** weights each of the terms of **X**. This pairwise product version of the dot product is used in the actual computations of dot products in the software routines that implement the methods being described.

While decision surfaces need not be linear, their simplicity when linear is appealing. Additionally, it can be shown that more complex decision surfaces can be implemented by linear decision surfaces preceded by appropriate preprocessing. That is, the space can be warped just as easily as the decision surfaces.

Two different approaches to the development of discriminants have been developed, *parametric* and *nonparametric*. Parametric methods of pattern recognition attempt to find classification surfaces or clustering definitions based on the statistical properties of the members of one or both classes of points. For example, Bayesian discriminants are developed using the mean vectors for the members of the classes and the covariance matrices for the classes. If the statistical properties of the classes cannot be calculated or estimated, then nonparametric methods can be used. Nonparametric methods attempt to find clustering definitions or classifiction surfaces by using the data themselves directly, without computing statistical properties. Examples of nonparametric methods include error correction feedback linear learning machines (perceptrons), simplex optimization methods of searching for separating classification surfaces, and the *K*-nearest-neighbor method.

Parametric Discriminant Development. The parametric pattern recognition methods use the mean vectors and covariance matrices (or other statistical measures) of the two classes into which the patterns fall as their basis for development of discriminants (Tou and Gonzalez 1974).

One parametric approach implements a quadratic discriminant function using the Bayes theorem. The equation expressing these discriminants is

$$d_k = \ln p_x - \ln C_k - (X - m_k)^T C_k^{-1} (X - m_k) \qquad k = 1, 2 \qquad (11)$$

where p_k is the a priori probability for class k, C_k is the covariance matrix for class k, and m_k is the mean vector for class k. This discriminant assumes a multivariate normal distribution of the data. A pattern X_i is put in the class for which $d_k(X_i)$ is greatest. Application of the Bayes quadratic discrimination method to the example data set yields a classification success rate of 99%, that is, only one pattern is classified incorrectly out of the 100.

If the assumption is made that the covariance matrices for the two classes are the same, the discriminant function developed using the Bayes theorem and the multivariate normal assumption simplifies from that above to

$$s = d_1(X) - d_2(X) = \ln p_1 - \ln p_2$$
$$+ X'X^{-1}(m_1 - m_2) - m_1'C^{-1}m_1 + m_2'C^{-1}m_2 \qquad (12)$$

where $s > 0$ for one class and $s < 0$ for the other class. The actual numerical computation can be done in any one of the three ways: using the covariance matrix for the entire data set for C, using the covariance matrix of class 1 for C, or using the covariance matrix for class 2 for C. These computations were all done with the example data set, and the success rates for classification were 96, 96, and 97%, respectively. A few more patterns were misclassified by these linear discriminants than by the quadratic one, but the success rate was still rather good.

Nonparametric Discriminant Development. The nonparametric pattern recognition programs develop their discriminants using the training set of patterns to be classified rather than statistical measures of their distributions.

The K-nearest-neighbor (KNN) method is conceptually the simplest. An unknown pattern is assigned to the class to which the majority of its nearest neighbors belong. The metric that is used to determine proximity is ordinarily the Euclidean metric, but any measure can be used. The classification decisions are based on the calculation of point-to-point distances.

The KNN method was applied to the example data set with the following results. Calculation of the distances from pattern 1 to each of the other 99 points is the first step in the classification of pattern 1. Then these distances are inspected, and it turns out that point 39 is nearest to point 1. The actual distance is computed as the square root of the sum of the squared differences between the autoscaled values of the descriptors (shown in Table 1, part E). The actual distance between points 1 and 39 is 1.055. The second closest point is number 47 at a distance of 1.435, the third closest is 31 at a distance of 1.767, and so on. The class of the first-nearest-neighbor point is class 1, so the first-nearest-neighbor method (1NN) would classify point 1 as being in class 1. The closest three points are all in class 1, so the 3NN method would also classify point 1 as being in class 1. When the procedure is applied to all 100 points, the results are as shown in Table 2. Approximately 90% of the patterns are correctly classified using the KNN method.

Another widely used nonparametric method is the linear learning machine or perception (Tou and Gonzalez 1974, Nilsson 1965). The algorithm for development of the discriminant is heuristic. A decision surface is initialized either arbitrarily or using the result from another linear discriminant development routine. Then the learning machine classifies one member of the training set at a time. When the current discriminant function correctly classifies a pattern, the discriminant is left unchanged. However, whenever an incorrect classification is made, the discriminant is altered in such a way that the error just committed is eliminated. The learning machine continues to classify the members of the training set repeatedly until no errors are committed or until the routine is externally terminated. This algorithm for the development of a linear discriminant function has the desirable property that if a training set of patterns is separable into the two classes, that is, if a solution exists, this method will find a solution. This existence proof enormously enhances the attractiveness of the linear learning machine.

The dot product of the weight vector and a pattern vector gives a scalar whose sign indicates on which side of the decision surface the pattern point lies:

$$\mathbf{WX} = s \tag{13}$$

TABLE 2. Results from K-Nearest-Neighbor Classifications

	Total	Correct	Incorrect	Percentage
First-Nearest-Neighbor Classification (1NN)				
Class 1	50	44	6	88
Class 2	50	47	3	94
Total	100	91	9	91
Three-Nearest-Neighbor Classification (3NN)				
Class 1	50	42	8	84
Class 2	50	48	2	96
Total	100	90	10	90

(An arbitrary decision must be made as to which subset of the data is to be called the positive class and which is to be called the negative class.) When pattern i of the training set is misclassified,

$$\mathbf{W}\mathbf{X}_i = s \tag{14}$$

in which s has the incorrect sign for classifying $\mathbf{X}_i$. The object is to calculate an improved weight vector $\mathbf{W}'$ such that

$$\mathbf{W}'\mathbf{X}_i = s' \tag{15}$$

where the sign of the scalar result s' is opposite what it was previously. The new weight vector is calculated from the old one by adding an appropriate multiple of s to it:

$$\mathbf{W}' = \mathbf{W} + c\mathbf{X}_i \tag{16}$$

Combining these equations gives

$$s' = \mathbf{W}'\mathbf{X}_i = (\mathbf{W} + c\mathbf{X}_i)\mathbf{X}_i \tag{17}$$

which can be algebraically rearranged to give

$$c = \frac{s' - s}{\mathbf{X}_i\mathbf{X}_i} \tag{18}$$

It remains only to choose a value for s' to complete the derivation. An effective method is to let $s' = -s$. This moves the decision surface so that after the feedback correction the point being classified, $\mathbf{X}_i$, is the same distance on the correct side of the new decision surface as it was previously on the incorrect side of the old decision surface. If $s' = -s$ is put into the equation,

$$c = -\frac{2s}{\mathbf{X}_i\mathbf{X}_i} \tag{19}$$

and $\mathbf{W}'$ can be calculated directly using the equations.

The training procedure involves iterating over all the pattern points in the training set and correcting the weight vector whenever an error is committed until the discriminant function converges on one that correctly classifies all the points. The process is called "learning" because the decision surface improves its performance at the classification task as its experience increases.

As mentioned above, this error correction procedure can be shown to find a solution if one exists. Therefore, the weight vector can be initialized arbitrarily, although it is obviously better practice to use whatever information is available to estimate a starting weight vector.

The question of training set size, N, versus the number of components per pattern, d, is important. While the exact ratio necessary for obtaining valid results cannot be calculated, it is true that the ratio N/d should be as large as posible. Values of N/d greater than approximately 3 are acceptable. This question has been discussed elsewhere (Tou and Gonzalez 1974) in some detail.

A linear learning machine program (program LM) is included in this chapter along with the set of example data and the results obtained during development of a decision surface separating the data set into its two classes.

Another nonparametric method develops a linear discriminant function through an iterative least-squares approach. The following error function is minimized:

$$Q = \sum_{i=1}^{m} [Y_i - F(s_i)]^2 \qquad (20)$$

where m is the total number of patterns in the training set, $Y_i = +1$ for one pattern class and $Y_i = -1$ for the other class, s_i is the dot product of the weight vector (discriminant) with the ith pattern, and $F(s_i)$ is the hyperbolic tangent funtion of s_i. The hyperbolic tangent function has the value of $+1$ for all positive values of s_i and -1 for all negative values of s_i. Therefore, minimizing Q is equivalent to minimizing the number of incorrect classifications. The function is nonlinear in the independent variables and thus cannot be solved directly, so an iterative algorithm is employed. This routine has been found to be particularly useful for dealing with data that are not completely separable into the two classes.

The search for a good discriminant that separates two clusters of points in an n-dimensional space from one another can be formulated as a linear programming problem (Ritter et al. 1975). The objective function whose value is to be minimized can be defined as the fraction of the training set of patterns that are incorrectly classified. Because two different discriminants can easily have the same classification power, it has been suggested that a secondary objective function be defined as the sum of the distances from the discriminant plane to the misclassified points. The secondary objective function is only invoked if the primary function is equal for two different discriminants being compared.

Clustering Methods

A subset of pattern recognition methods consists of clustering techniques that attempt to determine structural characteristics of a set of data by organizing the data into subgroups, clusters, or hierarchies. These methods are known as "unsupervised" because they attempt to define clusters solely on the basis of criteria derived from the data themselves (as opposed to finding clusters after knowing which data points belong to which categories.) The most widely used clustering method is hierarchical clustering. This method works by measuring the distances between all pairs of points, identifying the closest pair, combining them into a single point at the midpoint, recalculating the distance from this new point to every other point in the data set, finding the nearest pair, combining them, and so on. The process is continued until all the points have been combined. The resulting structure can be displayed as a dendrogram, which shows at what degree of similarity each pair of points was combined. The overall purpose is usually to simplify a data matrix that is too extensive for direct analysis. The product of hierarchical clustering analysis is a dendrogram or treelike structure that shows visually the hierarchy of similarity relationships in the original data. The techniques are independent of the nature of the data being classified, and they have been applied to a wide variety of problems drawn from engineering, the social sciences, and physical sciences, medicine, and biology (e.g., Hartigan 1975). Clustering routines must be considered to be an exploratory tool, and the absolute validity of dendrograms generated may be less important than the insights and suggestions generated regarding the data structure.

Modeling Methods

A pattern recognition method has been developed by Wold and coworkers called SIMCA (Wold and Sjostrom 1977). SIMCA is a set of procedures with which each class is separately modeled with a principal components model. Unknowns to be classified are fitted to the class models, and the classifications are made according to the statistics of the fits. An unknown pattern is assigned a probability for each class; if all probabilities are low, the pattern may be an outlier belonging to none of the original classes.

Program LM

Program LM and the subroutines TRAIN and PRED implement the error correction feedback linear learning machine algorithm for development of a linear binary pattern classifier. The mathematics involved in the algorithm has been presented in the section on nonparametric discriminant development. The main program, LM, inputs a set of data, randomly chooses a training set, initializes a number of parameters, calls the training routine, and calls the prediction routine.

The program presented here is similar to a program previously published (Jurs and Isenhour 1975), but this version has been changed in some ways. It has been made fully interactive, with the output all being sent to device 1. The data set to be analyzed is taken from file 5. All other parameters are initialized inside the routine for simplicity. Of course, all the parameters could be input by the user upon request through a series of READ statements if desired.

The array named DATA contains the raw data or patterns. It is dimensioned for up to 100 patterns of up to five descriptors per pattern. The data are input from an input file named INPUT, which contains card images of the data. No normalization of the data is performed by LM. The variable named LIST contains the category of each pattern, with +1 for one class and −1 for the other class. This information is input along with the data themselves. A training set of 80 patterns is chosen randomly, and the remaining 20 patterns are put in the prediction set. The weight vector is initialized so that each component has a value of zero except for the $(d + 1)$st component with a value of unity. NTRSET is the number of patterns forming the training set, here 80. NUM is the number of descriptors per pattern, here 5. NPASS contains the number of error correction feedbacks that will be allowed before the learning routine is terminated due to lack of convergence, here 1000. TSHD is the nonzero threshold that is to be used during training, here 0.75. This has the effect of giving the decision surface being developed a thickness. NCONV is a flag variable that will report back to the calling program whether convergence was attained or not. IDTR is an array that contains the pattern numbers of the patterns forming the training set.

Subroutine TRAIN implements the error correction training procedure for seeking a binary pattern classifier. To be correctly classified, a point must not only be on the proper side of the decision surface but also be outside of the actual volume occupied by the decision surface. The subroutine uses a subsetting procedure to attempt to make progress in the overall task as efficiently as possible. On the first pass through the data set, the entire training set is used with feedback corrections being made as necessary. At the same time, the routine keeps track of which members of the training were set misclassified. This information is stored in variable NSS. On the second pass, only those patterns missed on the first pass are classified, and a third subset is constructed. The sequence repeats until no patterns are misclassified, that is, until the subset is empty. Then the entire training set is classified, and the entire sequence begins again. The training program prints out the number of patterns misclassified on subsequent passes through the subsetting procedure. Only when two zeros appear in sequence is classification of the training set perfect. Subroutine TRAIN terminates when all the members of the training set are classified correctly or the allowed number of feedbacks is exceeded. Then it prints out the weight vector and the number of feedbacks employed during the training.

Subroutine PRED uses the trained weight vector to predict the classes of the members of the prediction set. Two calls to PRED are included in LM, one with the threshold value of 0.75 and one with the threshold value set to zero. PRED prints out the results of the predictions.

Program LM was executed with the example data set, and the output is presented. The identities of the 20 patterns chosen for the prediction set are output. The training routine outputs the number of patterns misclassified during each pass through either the entire training set or the subsets as a means to monitor progress during training. When two zeros appear, this means that training is complete. Then the weight vector is output, and the total number of feedbacks is printed. The prediction routine is called twice. With the threshold set to 0.75, 3 of the 20 patterns in the prediction set were not predicted because they were found to fall within the volume occupied by the decision surface. All of the remaining 17 patterns were classified correctly. With the threshold set to zero, all 20 patterns were correctly classified.

```
      PROGRAM LM
C....
C.... BASIC LEARNING MACHINE PROGRAM
C....
      DIMENSION DATA(5,100),W(6),LIST(100),IDTR(100),IDPR(100)
     X   ,ITEMP(100)
      COMMON /IOUNIT/ NINP,NOUT
      OPEN (5,FILE='INPUT')
      NINP = 5
      NOUT = 1
      WRITE (NOUT,1)
    1 FORMAT (' LEARNING MACHINE PROGRAM',/)
C.... INITIALIZE RANDOM NUMBER GENERATOR
      IX = 521
      RR = RAND(IX)
C.... SET VALUES OF PARAMETERS
      NTRSET = 80
      NPRSET = 20
      TSHD = 0.75
      NTOT = NTRSET + NPRSET
      NPASS = 1000
      NUM = 5
C.... INPUT DATA SET
      DO 10 I=1,NTOT
   10 READ (NINP,11) LIST(I),(DATA(J,I),J=1,NUM)
   11 FORMAT (I5,5F10.3)
C.... RANDOMLY CHOOSE TRAINING SET
      DO 20 I=1,NTOT
   20 ITEMP(I) = I
      DO 30 I=1,NTRSET
   25 ITEST = 100.0*RAND(IX)
      IF (ITEMP(ITEST).EQ.0) GO TO 25
      IDTR(I) = ITEST
      ITEMP(ITEST) = 0
   30 CONTINUE
C.... REMAINING PATTERNS FORM PREDICTION SET
      II = 0
      DO 40 I=1,NTOT
      IF (ITEMP(I).EQ.0) GO TO 40
      II = II + 1
      IDPR(II) = I
   40 CONTINUE
      WRITE (NOUT,41) (IDPR(I),I=1,NPRSET)
   41 FORMAT (/,' PREDICTION SET MEMBERS',/,(' ',5I5))
```

```
C.... INITIALIZE WEIGHT VECTOR
      DO 50 J=1,NUM
   50 W(J) = 0.0
      W(NUM+1) = 1.0
      CALL TRAIN (DATA,W,LIST,NTRSET,NUM,NPASS,TSHD,NCONV,IDTR)
C.... CALL PREDICTION ROUTINE WITH THRESHOLD OF 0.75
      CALL PRED (DATA,LIST,W,NUM,TSHD,NPRSET,IDPR)
C.... CALL PREDICTION ROUTINE WITH THRESHOLD OF 0.0
      TSHD = 0.0
      CALL PRED (DATA,LIST,W,NUM,TSHD,NPRSET,IDPR)
      STOP
      END
C-------------------------------------------------------------
      SUBROUTINE TRAIN (DATA,W,LIST,NTRSET,NUM,NPASS,TSHD,NCONV,IDTR)
C....
C.... IMPLEMENTS ERROR-CORRECTION FEEDBACK LINEAR LEARNING MACHINE
C....
      DIMENSION DATA(5,100),W(6),NSS(100),KPNT(20),LIST(100),IDTR(100)
      COMMON /IOUNIT/ NINP,NOUT
      NCONV = 0
      WRITE (NOUT,2)
    2 FORMAT (/,10X,'TRAINING ROUTINE')
      NUMM = NUM + 1
      NF = 0
      KNK = 0
      KNV = 0
C.... START OF MAIN LOOP (RETURN FROM STMT 206)
   51 KKK = 0
      IF (KNV) 54,54,53
   53 NDSS = KNV
      GO TO 65
   54 NDSS = NTRSET
      DO 60 I=1,NTRSET
   60 NSS(I) = IDTR(I)
C.... THE 200 LOOP CLASSIFIES THE NDSS MEMBERS OF THE CURRENT SUBSET
   65 DO 200 IR=1,NDSS
      I = NSS(IR)
C.... THE 70 LOOP CALCULATES THE DOT PRODUCT BETWEEN THE PATTERN
C....       AND THE CURRENT WEIGHT VECTOR
      S = W(NUMM)
      DO 70 J=1,NUM
   70 S = S + DATA(J,I)*W(J)
C.... NEXT THREE STMTS TEST FOR THE CORRECT ANSWER
      IF (LIST(I)) 95,95,96
   95 IF (S+TSHD) 200,200,116
   96 IF (S-TSHD) 115,115,200
C.... CALCULATE THE CORRECTION INCREMENT C
  115 C = 2.0*(TSHD-S)
      GO TO 117
  116 C = 2.0*(-TSHD-S)
  117 XX = 1.0
      DO 120 J=1,NUM
  120 XX = XX + DATA(J,I)**2
      C = C/XX
C.... THE 130 LOOP CORRECTS THE WEIGHT VECTOR
      DO 130 J=1,NUM
  130 W(J) = W(J) + C*DATA(J,I)
      W(NUMM) = W(NUMM) + C
      KKK = KKK + 1
      NSS(KKK) = I
      NF = NF + 1
  200 CONTINUE
      KNV = KKK
      KNK = KNK + 1
      KPNT(KNK) = KNV
      IF (KNK - 20) 205,203,203
  203 WRITE (NOUT,204) KPNT
```

```
    204 FORMAT (' ',20I3)
        KNK = 0
C.... STMT 205 TESTS FOR EXCESS NUMBER OF FEEDBACKS
    205 IF (NF-NPASS) 206,211,211
C.... STMT 206 TESTS WHETHER CURRENT SUBSET IS ENTIRE TRAINING SET
    206 IF (NDSS-NTRSET) 51,207,51
C.... STMT 207 TESTS WHETHER ZERO ERRORS WERE COMMITTED
    207 IF (KNV) 51,212,51
    211 NCONV = 1
C.... SUMMARY OUTPUT OF TRAINING ROUTINE
    212 IF (KNK.GT.0) WRITE (NOUT,204) (KPNT(K),K=1,KNK)
        WRITE (NOUT,213) (W(J),J=1,NUMM)
    213 FORMAT (/,10X,'WEIGHT VECTOR',/,(' ',F17.3))
        WRITE (NOUT,214) NF
    214 FORMAT (/,10X,'FEEDBACKS',I6)
        RETURN
        END
C-------------------------------------------------------------
        SUBROUTINE PRED (DATA,LIST,W,NUM,TSHD,NPRSET,IDPR)
C....
C.... PREDICTION ROUTINE
C....
        DIMENSION DATA(5,100),W(6),LIST(100),IDPR(100)
        COMMON /IOUNIT/ NINP,NOUT
        LW1 = 0
        LW2 = 0
        KW = 0
        NPA = 0
        NNA = 0
        DO 120 II=1,NPRSET
        I = IDPR(II)
        S = W(NUM+1)
        DO 50 J=1,NUM
     50 S = S + DATA(J,I)*W(J)
        IF (ABS(S)-TSHD) 101,102,102
    101 KW = KW + 1
        GO TO 120
    102 IF (LIST(I)) 103,103,105
    103 NNA = NNA + 1
        IF (-S-TSHD) 104,104,120
    104 LW1 = LW1 + 1
        GO TO 120
    105 NPA = NPA + 1
        IF (S-THSD) 106,106,120
    106 LW2 = LW2 + 1
    120 CONTINUE
        WRITE (NOUT,121) TSHD
    121 FORMAT (///,' PREDICTION WITH THRESHOLD =',F7.2)
        LWT = LW1 + LW2
        JW = NPA + NNA
        PW = 100.0-FLOAT(LWT)/FLOAT(JW)*100.0
        PW1 = 100.0-FLOAT(LW1)/FLOAT(NNA)*100.0
        PW2 = 100.0-FLOAT(LW2)/FLOAT(NPA)*100.0
        WRITE (NOUT,122) JW,KW,LWT
    122 FORMAT (/,I10,'   NUMBER PREDICTED',/,I10,
       X        '   NUMBER NOT PREDICTED',/,I10,
       X        '   NUMBER PREDICTED INCORRECTLY')
        WRITE (NOUT,123)
    123 FORMAT (/,12X,'OVERALL',13X,'NO CLASS',13X,'YES CLASS')
        WRITE (NOUT,124) LWT,JW,PW,LW1,NNA,PW1,LW2,NPA,PW2
    124 FORMAT (3(I10,'/',I3,1X,F6.2))
        RETURN
        END
C-------------------------------------------------------------
        FUNCTION RAND (IX)
C.... PORTABLE FORTRAN RANDOM NUMBER GENERATOR
C....    FROM A.C.M. TRANS. MATH. SOFTWARE, 5, #2, 132 (1979)
```

```
C....      BY LINUS SCHRAGE
C....      USES THE RECURSION   IX = IX * A (MOD P)
C....
C.... INITIALIZE WITH SEED 0 < SEED < 2**31-1
C....
C.... USE EITHER RAND:  0 < RAND < 1
C....           OR IX:  0 < IX < 2**31 -1
C....
C.... CHECKING VALUES: IF IX(0)=1, THEN IX(1000)=522329230
C....
C.... IX IN CALLING LIST MUST BE INTEGER*4 IN CALLING PROGRAM
C....
           INTEGER A, P, IX, B15, B16, XHI, XALO, LEFTLO, FHI, K
C....   7**5, 2**15, 2**16, 2**31-1
           DATA A/16807/, B15/32768/, B16/65536/, P/2147483647/
C....
C.... GET 15 HIGH ORDER BITS OF IX
           XHI = IX/B16
C.... GET 16 LOW BITS OF IX AND FORM LO PRODUCT
           XALO = (IX-XHI*B16)*A
C.... GET 15 HIGH ORDER BITS OF LO PRODUCT
           LEFTLO = XALO/B16
C.... FORM THE 31 HIGHEST BITS OF FULL PRODUCT
           FHI = XHI*A + LEFTLO
C.... GET OVERFLOW PAST 31ST BIT OF FULL PRODUCT
           K = FHI/B15
C.... ASSEMBLE ALL THE PARTS AND PRESUBTRACT P
           IX = (((XALO-LEFTLO*B16) - P) + (FHI-K*B15)*B16) + K
C.... ADD P BACK IN IF NECESSARY
           IF (IX.LT.0) IX = IX + P
C.... MULTIPLY BY 1/(2**31-1)
           RAND = FLOAT (IX) * 4.656612875E-10
           RETURN
           END
```

LEARNING MACHINE PROGRAM

PREDICTION SET MEMBERS

2	3	11	14	15
16	19	20	28	38
39	45	46	48	58
62	65	77	79	100

TRAINING ROUTINE

```
29 14  8  5  3  2  2  2  2  2  1  0 10  4  4  2  1  0  6
 3  3  1  0  8  5  1  0  9  5  5  5  4  4  4  4  4  4  4
 4  4  3  3  3  3  3  3  2  2  2  2  0  8  3  3  2  1  0  7
 4  4  3  2  2  0  6  5  4  4  4  4  4  4  3  2  1  0  4  2
 2  2  0  6  3  3  0  3  0  7  7  7  6  6  6  6  6  6  6  6
 6  6  5  5  5  5  5  5  5  5  5  5  5  5  5  5  5  5  5  5
 4  4  4  4  4  4  4  3  3  3  3  3  2  1  0  0
```

WEIGHT VECTOR

```
    0.573
    1.187
    0.605
   -0.971
   -0.544
    0.902
```

FEEDBACKS   532

```
PREDICTION WITH THRESHOLD =    0. 75

    17   NUMBER PREDICTED
     3   NUMBER NOT PREDICTED
     0   NUMBER PREDICTED INCORRECTLY

         OVERALL              NO CLASS              YES CLASS
     0/ 17 100. 00        0/  3 100. 00        0/ 14 100. 00


PREDICTION WITH THRESHOLD =    0. 00

    20   NUMBER PREDICTED
     0   NUMBER NOT PREDICTED
     0   NUMBER PREDICTED INCORRECTLY

         OVERALL              NO CLASS              YES CLASS
     0/ 20 100. 00        0/  6 100. 00        0/ 14 100. 00
```

14.2 SELECTED CHEMICAL APPLICATIONS OF PATTERN RECOGNITION

Application studies of chemical problems using pattern recognition techniques have been reported in a number of areas (Jurs and Isenhour 1975, Varmuza 1980, Kowalski 1980, Kryger 1981, Kowalski and Wold 1982, Delaney 1984). The book by Jurs and Isenhour (1975) contains 113 references in all, of which at least 60 refer to chemical problems. The book by Varmuza (1980) contains 435 references, the reviews by Kowalski (1980, Kowalski and Wold 1982) contain many pattern recognition application references, the review article by Kryger (1981) contains 130 references, and the review by Delaney (1984) contains 44 references on pattern recognition. Thus, there is a wealth of primary literature dealing with chemical applications of pattern recognition. This section will summarize some of the areas of application and provide references to a sampling of the primary publications. The best way for an interested reader to sample the information available in this area would be to consult one of the reviews.

Spectral Data Analysis

The elucidation of chemical structure information from spectral data is a long-standing problem of chemistry. This is the area first studied and most intensively studied using pattern recognition. Studies have been done with mass spectra, infrared spectra, Raman spectra, electrochemical data, gamma ray spectra, proton and ^{13}C NMR spectra, and Auger spectra.

Mass Spectrometry. The analysis of low-resolution mass spectral data to obtain structural information was the first chemical problem studied by pattern recognition methods (Raznikov and Talroze 1966). Papers dealing with mass spectrometric analysis are numerous enough that Varmuza (1980) chose to list 14 review papers on the topic. The majority of the work has dealt with the recognition of molecular substructures or functional groups from the mass spectra of compounds. Some example studies are as follows: sequence analysis of oligodeoxyribonucleotides (Burgard et al. 1977), interpretation of steroid mass spectra (Rotter and Varmuza 1978), determination of molecular structure parameters (Jurs et al. 1970).

Infrared Spectra. The identification of function groups from the infrared spectra of organic compounds has been reported (e.g., Woodruff and Munk 1977). The infrared spectra are ordinarily digitized into equally spaced subintervals in order to represent the spectra as pattern vectors. Binary-coded spectra with pattern vector component restricted to 1 or 0 have also been studied.

NMR Spectra. Both proton and ^{13}C NMR spectra have been studied, and one review has appeared (Wilkins 1978). Example studies include the influence of substituent effects on NMR patterns (Edlund and Wold 1980) and separation of structural classes (Brunner et al. 1975).

Classification of Complex Mixtures

Materials or mixtures can be classified into categories (e.g., origin) by pattern recognition. Examples from a number of diverse areas can be found in the literature: manufacturers and grades of paper (Duewer and Kowalski 1975), quarry sites of archaeological artifacts (McGill and Kowalski 1977), sources of atmospheric particulate matter (Gaarenstroom, et al. 1977), classification of wines (Kwan and Kowalski 1980), determination of olive oil origin (Forina and Tiscornia 1982), identification of crude oil samples (Clark and Jurs 1979), determination of the clinical status of patients from urine samples (Rhodes et al. 1981), and classification of cancer cells (Jellum et al. 1981).

Prediction of Properties from Molecular Structure

A number of studies of the application of pattern recognition to the problem of searching for relationships between molecular structure and biological activity have been reported. A large fraction of this type of research is involved with the generation of appropriate descriptors from the molecular structures available. Areas of study that have been reported in the literature include drug structure–activity relations (SAR), studies of chemical communicants, studies of structure–toxicity relations, and others. Early applications of pattern recognition to drug design have been reviewed by Kirschner and Kowalski (1979). This area of application of pattern recognition is briefly mentioned by Varmuza (1980). A book describing one approach to SAR research has appeared (Stuper et al. 1979).

A few representative SAR studies are as follows: a study of 200 drugs for anticancer activity (Kowalski and Bender 1974), a study of 9-anilino-acridines for antitumor selectivity (Henry et al. 1982), studies of drugs of accepted therapeutic value (Menon and Cammarata 1977), structure–carcinogenicity potential (Rose and Jurs 1982), olfactory quality or organic compounds (Jurs et al. 1981), and structure–carcinogenic potential of PAH (Norden and Wold 1978).

REFERENCES

T. R. Brunner, C. L. Wilkins, R. C. Williams, and P. J. McCombie, "Pattern Recognition Analysis of Carbon-13 Free Induction Decay Data," *Anal. Chem.*, **47**, 662–665 (1975).

D. R. Burgard, S. P. Perone, J. L. Wiebers, "Sequence Analysis of Oligodeoxy-ribonucleotides by Mass Spectrometry. 2. Application of Computerized Pattern Recognition to Sequence Determination of Di-, Tri-, and Tetranucleotides," *Biochemistry*, **16**, 1051–1057 (1977).

H. A. Clark and P. C. Jurs, "Classification of Crude Oil Gas Chromatograms by Pattern Recognition Techniques," *Anal. Chem.*, **51**, 616–623 (1979).

M. F. Delaney, "Chemometrics," *Anal. Chem.*, **56**, 261R–277R (1984).

D. L. Duewer and B. R. Kowalski, "Forensic Data Analysis by Pattern Recognition. Categorization of White Bond Papers by Elemental Composition," *Anal. Chem.*, **47**, 526–530 (1975).

U. Edlund and S. Wold, "Interpretation of NMR Substituent Parameters by the Use of a Pattern Recognition Approach," *J. Mag. Res.*, **37**, 183 (1980).

M. Forina and E. Tiscornia, "Pattern Recognition Methods in the Prediction of Italian Olive Oil Origin by Their Fatty Acid Content," *Ann. Chim.*, **72**, 143–155 (1982).

I. E. Frank and B. R. Kowalski, "Chemometrics," *Anal. Chem.*, **54**, 232R–243R (1982).

P. D. Gaarenstroom, S. P. Perone, and J. L. Moyers, "Application of Pattern Recognition and Factor Analysis for Characterization of Atmospheric Particulate Composition in Southwest Desert Atmosphere," *Environ. Sci. Tech.*, **11**, 795–800 (1977).

J. A. Hartigan, *Clustering Algorithms*, Wiley, New York, 1975.

D. R. Henry, P. C. Jurs, and W. A. Denny, "Structure-Antitumor Activity Relationships of 9-Anilinoacridines Using Pattern Recognitions," *J. Med. Chem.*, **25**, 899–908 (1982).

E. Jellum, I. Bjoernson, R. Nesbakken, E. Johansson, and S. Wold, "Classification of Human Cancer Cells by Means of Capillary Gas Chromatography and Pattern Recognition Analysis," *J. Chromatogr.*, **217**, 231–237 (1981).

P. C. Jurs, and T. L. Isenhour, *Chemical Applications of Pattern Recognition*, Wiley-Interscience, New York, 1975.

P. C. Jurs, C. L. Ham, and W. E. Brugger, "Computer Assisted Studies of Chemical Structure and Olfactory Quality Using Pattern Recognition Techniques," *A.C.S. Symp. Ser.*, **148**, 143 (1981).

P. C. Jurs, B. R. Kowalski, T. L. Isenhour, and C. N. Reilley, "Computerized Learning Machines Applied to Chemical Problems. Molecular Structure Parameters from Low Resolution Mass Spectrometry," *Anal. Chem.*, **42**, 1387–1294 (1970).

G. L. Kirschner and B. R. Kowalski, "The Application of Pattern Recognition to Drug Design," in *Drug Design*, Vol. 8, E. J. Ariens (Ed.), Academic, New York, 1979.

B. R. Kowalski, "Chemometrics," *Anal. Chem.*, **52**, 112R–122R (1980).

B. R. Kowalski and C. F. Bender, "The Application of Pattern Recognition to Screening Prospective Anticancer Drugs. Adenocarcinoma 755 Biological Activity Test," *J. Amer. Chem. Soc.*, **96**, 916–918 (1974).

B. R. Kowalski and S. Wold, "Pattern Recognition in Chemistry," in *Handbook of Statistics*, Vol. 2, P. R. Krishnaiah and L. N. Kanal (Eds.), North-Holland, New York, 1982.

L. Kryger, "Interpretation of Analytical Chemistry Information by Pattern Recognition Methods: A Survey," *Talanta*, **28**, 871–887 (1981).

W. O. Kwan and B. R. Kowalski, "Correlation of Objective Chemical Measurements and Subjective Sensory Evaluations. Wines of *Vitis Vinifera* Variety 'Pinot Noir' from France and the United States," *Anal. Chim. Acta*, **122**, 215–222 (1980).

J. R. McGill and B. R. Kowalski, "Recognizing Patterns in Trace Elements," *Appl. Spectrom.*, **31**, 87–95 (1977).

G. K. Menon and A. Cammarata, "Pattern Recognition II: Investigation of Structure–Activity Relationships," *J. Pharm. Sci.*, **66**, 304–314 (1977).

B. Norden, U. Edlund, and S. Wold, "Carcinogenicity of Polycyclic Aromatic Hydrocarbons Studied by SIMCA Pattern Recognition," *Acta Chem. Scand.*, **B32**, 602–608 (1978).

V. V. Raznikov and V. L. Talroze, *Dokl. Akad. Nauk SSSR*, **170**, 379 (1966).

G. Rhodes, M. Miller, M. L. McConnell, and M. Novotny, "Metabolic Abnormalities Associated with Diabetes Mellitus, as Investigated by Gas Chromatography and Pattern Recognition Analysis of Profiles of Volatile Metabolites," *Clin. Chem.*, **27**, 580–585 (1981).

G. L. Ritter, S. R. Lowry, C. L. Wilkins, and T. L. Isenhour, "Simplex Pattern Recognition," *Anal. Chem.*, **47**, 1951 (1975).

S. L. Rose and P. C. Jurs, "Computer-Assisted Studies of Structure Activity Relationships of N-Nitroso Compounds Using Pattern Recognition," *J. Med. Chem.*, **25**, 769–776 (1982).

H. Rotter and K. Varmuza, "Computer-Aided Interpretation of Steroid Mass Spectra by Pattern Recognition Methods. Part III. Computation of Binary Classifiers by Linear Regression," *Anal. Chim. Acta*, **103**, 61–71 (1978).

A. J. Stuper, W. E. Brugger, and P. C. Jurs, *Computer Assisted Studies of Chemical Structure and Biological Function*, Wiley-Interscience, New York, 1979.

J. T. Tou and R. C. Gonzalez, *Pattern Recognition Principles*, Addison-Wesley, Reading, MA, 1974.

K. Varmuza, *Pattern Recognition in Chemistry*, Springer-Verlag, Berlin, 1980.

K. Varmuza, "Some Aspects of the Application of Pattern Recognition Methods in Chemistry," in *Computer Applications in Chemistry*, S. R. Heller and R. Potenzone (Eds.), Elsevier Scientific, Amsterdam, 1983.

P. H. Wang (Ed.), *Graphical Representation of Multivariate Data*, Academic, New York, 1978.

C. L. Wilkins and P. C. Jurs, "Fourier and Hadamard Transforms in Pattern Recognition," in *Transform Techniques in Chemistry*, P. R.; Griffiths (Ed.), Plenum, New York, 1978, pp. 307–331.

S. Wold and Sjostrom, "A Method for Analyzing Chemical Data in Terms of Similarity and Analogy," *A.C.S. Symp. Ser.*, **52**, 243–282 (1977).

H. B. Woodruff and M. E. Munk, "Computer-Assisted Interpretation of Infrared Spectra," *Anal. Chim. Acta*, **95**, 13–23 (1977).

15

ARTIFICIAL INTELLIGENCE AND EXPERT SYSTEMS

Artificial intelligence (AI) is a field of research in which the following goals are pursued:

1. To understand the nature of intelligence using the paradigm of information processing.
2. To construct theories and build systems through programming computers that exhibit intelligent behavior requiring reasoning and perception.
3. To make computers more useful by making them exhibit intelligent behavior.

The central activity involved in intelligent behavior is information processing. Intelligence in human behavior is associated with tasks such as understanding natural language, learning, reasoning, solving problems, and playing complex games such as chess or go. All of these areas have been studied within AI. Problem-solving programs have been developed in chemistry, biology, geology, engineering, and medicine that can perform tasks at a level similar to that of human experts. AI is practiced as an experimental science, that is, the construction of computer programs and demonstration of their capabilities are considered to be a vital part of the research.

Artificial intelligence has been a field of research for only about 30 years. Its underpinnings were advances in mathematical logic during the 1930s and 1940s and advances in computational capability during the 1940s and early 1950s. The development of electronic digital computers made possible the implementation of tasks that involved intelligence for the first time. An early landmark paper was by Turing (1950) in which he addressed many of the fundamental questions of intelligent behavior of machines and proposed what has come to be called the *Turing test*. Researchers were able to build systems to test their theories of intelligent behavior, leading to the typical scientific cycle of theory, experiment, theory revision, and so on. Progress was rapid. The early research focused on game-playing programs as a domain for study. Soon other domains were studied, including theorem proving, language understanding, vision, speech recognition, and so on. In the 1980s, AI is a rapidly growing field with all the hallmarks of a relatively mature scientific discipline: a professional society of its own (American Association for Artificial Intelligence), both scientific and popular journals, textbooks, university courses, proponents, opponents, and so on (Duda and Shortliffe, 1983).

There is a deeply rooted *paradox* in AI. On the one hand, computers are completely inflexible, they do slavishly what they are told to do, and they never deviate from following instructions exactly. On the other hand, the hallmark of intelligence is flexibility, adaptability, and the ability to respond or decide as the situation suggests. The conciliation of computer characteristics with the demands of intelligent behavior is an accomplishment and a goal of AI research. This aspect of AI has been discussed at length by Simon (1984) and Hofstadter (1979).

Intelligent behavior is developed in computers through implementation of software, that is, the programs. The programs execute on computer hardware, but that fact is relatively unimportant. It is the symbol manipulation done by the programs that is relevant, not the details of how the electronic circuits accomplish it.

There are two basic ingredients in intelligent behavior: search and knowledge. Search refers to the ability to create a space of possibilities that is large enough to contain the sought solution to the problem and then searching for that solution. Often the spaces for real problems increase in size very quickly as a function of the size of the problem. This is true, for example, in the looking ahead in a chess game for possible moves. There are something like 10^{120} sequences of legal moves in a chess game. It is beyond the power of any computer to examine them all and thereby exhaustively generate the perfect response move in chess. Such combinatorial explosions of space size often form a fundamental limitation on the capabilities of intelligent programs.

The second basic ingredient of intelligent behavior is the possession of knowledge. A popular phrase of recent AI research is "Knowledge is power." AI applications make great demands on the knowledge bases employed. The knowledge is usually diverse and interrelated. The knowledge can be used to guide the search so that the number of possible solutions that must be examined is limited. The representation of the knowledge must be effective, so that the knowledge can be used by the system to attain its goals. The organization of the knowledge so that it is accessible and can be found when needed is a very important issue in AI research. Several different paradigms have been developed for the organization of the knowledge bases in AI programs. One is standard symbolic logic, such as "All fish swim." Another representation is that of production systems that use if–then rules, which will be described below. Specialized representations such as frames or scripts have been developed as well. In addition to the organization of the knowledge, practical AI systems must provide a means for acquiring the knowledge and inserting it into the knowledge base. The automation of knowledge acquisition is a research area within AI. The next step of knowledge acquisition, learning from experience or examples, is a newly emerging area of AI research.

An example based on an idea of Rich (1984) will illustrate the different demands made of knowledge bases by AI programs as opposed to ordinary programs. Take as an example of a normal data base the information contained in an ordinary almanac. Facts abound in such a data base, often presented in tabular form. This type of information can be stored and manipulated by computer very easily. Questions based on such facts can be answered by straightforward methods. The answers to questions of the following types can be looked up in the appropriate part of the data base: "What is the tallest mountain in the world?" or "Who was the first secretary general of the United Nations?"

Now, suppose that the following fact is contained in the almanac: Sacramento is the capital of California. Based on this isolated fact, a person would be able to answer any one of the following questions: "Is San Diego the capital of California?" "What state is Sacramento in?" "Where is the government of California?" "Is Sacramento in the United States?" The answers to these questions are not contained in the simple piece of factual information given. Additional information is needed to answer these questions, information such as: states have just one capital; a capital of a state is in that state; if a is in b, and b is in c, then a is in c; California is a state; a state has its government in its capital; the United States is comprised of states; and so on. The kind of knowledge manipulation required to put facts together in a flexible way is one key to AI program success.

One important class of AI programs consists of "expert systems" that are designed to serve as consultants for decision making. The fundamental aspect of intelligence that AI systems must implement is how to represent large amounts of information in a way that allows for its effective use. For a program to exhibit intelligence, it must have access to large amounts of knowledge and must know how to manipulate it and use it. The manipulations are sometimes general, but they are often specific to the domain of interest. This approach to the development of expert systems has been called the knowledge-based approach.

Expert systems have been developed within a number of narrow domains. The knowledge bases have been acquired from human experts, who are experts precisely because of their extensive knowledge within their domain. An important characteristic of expert systems is that they are not limited to the knowledge of one person but can store and use the rules as known to a group of community of experts. One characteristic of this knowledge that must be captured in an expert system is that the knowledge is not always certain or objective. It can be subjective, judgmental, or even rules of thumb. The type of information used and the type of processing that is done with it often does not conform to the definition of *algorithm*. Rather, the correct labeling is *heuristic*. A heuristic is a rule or way of proceeding that is usually all right but may not always be optimum. Several examples of heuristics are as follows: control of the center of the board is important in chess; a falling barometer means that it very well may rain soon; red sky at night is sailor's delight; a florid complexion may mean high blood pressure; and so on. Expert systems must deal with their knowledge base of information through the use of heuristics in attempting to make decisions.

The most common form for representation of information is that of production rules, or if–then rules. They have the form

$$\text{IF } \langle \text{premise} \rangle \text{ THEN } \langle \text{action} \rangle$$

where the conditions with ⟨premise⟩ can be very complicated and the ⟨action⟩ can also contain a series of subactions. Ordinarily, only human experts have the knowledge of the proper pairs of premises and actions.

The basic research issues within AI are knowledge acquisition, knowledge representation, inference and uncertainty, and explanation. The current bottleneck in the development of expert systems is the acquisition and representation of the knowledge base.

Knowledge Acquisition. Domain knowledge must be extracted from the human experts. In the development of current expert systems, this has largely been done manually by collaboration between an AI researcher and a domain expert. This has meant having computer scientists interview experts in the domain of applications. The knowledge acquired in this way usually is in the form of English sentences. It must be structured by the computer scientist so that it can be properly represented by the computer system. This task has come to be called *knowledge engineering*. The slow going of this manual method has led many to seek methods that would automate knowledge acquisition or, better yet, allow systems to learn from examples presented in a natural way. This will remain a goal for some time, however.

Knowledge Representation. Knowledge must be represented inside the expert system so that it can be used effectively. The representation must be accessible and flexible. The knowledge must be represented so that it can be modified and augmented, and the knowledge must be represented so that the expert system can explain its actions to humans.

Inference and Uncertainty. Decisions are made in expert systems by weighing evidence, taking into account the (possibly many) rules that have bearing on the question at hand, and sometimes calculating probabilities. However, the most effective methods for incorporating uncertainty in the validity of the rules or the probabilities of correct application as a function of the situation have not been fully defined as yet.

Explanation. This refers to the necessity for expert systems to be able to provide their reasoning sequence to the human user so that he can understand and critique it. This is necessary not so much for the proper functioning of the expert system as to improve the system's veracity with its users. Ability to explain actions and the bases for decisions is required of human experts, and it is necessary for AI expert systems also.

A number of specific AI systems have been developed and introduced into the commercial marketplace. Some noteworthy examples follow.

MACSYMA is a symbolic manipulation system that functions as a mathematical aid.

R1 is an expert system that aids in the configuration of Digital Equipment Corporation VAX computer systems. Bachant and McDermott (1984) state; "Given a customer's purchase order, R1 determines what, if any, substitutions and additions have to be made to the order to make it consistent, complete, and produce a number of diagrams showing the spacial and logical relationships among the 50 to 150 components that typically constitute a system."

MYCIN performs diagnosis for blood diseases due to bacterial infections and prescribes treatment.

CADUCEUS is an advisor in internal medicine.

PROSPECTOR is an expert geologist system that seeks commercially exploitable mineral deposits.

MOLGEN is an expert system for designing experiments in molecular biology.

DENDRAL is an expert system for elucidation of molecular structures from spectroscopic data.

The DENDRAL Project

One of the first expert systems to be developed, DENDRAL, is also one of the most widely used. It analyzes mass spectra to aid in organic structure elucidation. The project was begun by Lederberg in 1964 as he considered a long-standing problem of organic chemistry, namely, how many structures can be constructed using a fixed set of atoms? The program he developed to answer this question for simple, acyclic structures he termed the DENDRAL program for *dendritic algorithm*. The early reports on DENDRAL were a series of technical reports to NASA, which are summarized in Lederberg (1969). Later, the system also incorporated mass spectral evaluation to become what has been known as the DENDRAL project. A review is given by Lindsay et al. (1980).

The general problem of structure elucidation is approached by DENDRAL in three stages:

1. Planning by developing constraints.
2. Generation of structures using constraints.
3. Ranking and testing of structures.

Planning. The planning stage develops limitations on the structures that are feasible answers to the structure elucidation question being posed. To do this, the planner uses whatever information is input by the user regarding structural subunits, substructures, or "superatoms," that form constraints. These can either be structural units that must be present within the suggested solution structures or structural units that must be absent from the suggested structures. These constraints can sometimes be deduced from spectroscopic data, usually mass spectra. The more constraints that can be introduced to the problem, the fewer potential solution structures there will be and the easier the solution of the problem.

Testing. The testing is accomplished by developing predicted spectral data for each candidate solution structure and then comparing each predicted spectrum to the actual, observed spectrum of the unknown. The closest fits are ranked the highest and retained for further study.

Generation. The part of the DENDRAL system that works on the generation problem is now called CONGEN for constrained structure generation. CONGEN is the least tied to specific spectroscopic data and is of the greatest general utility. It has increased over the years in sophistication and practicality so that it is now routinely used for real problems.

In the mid-1970s, the Dendral group reported a program called meta-Dendral. This program dealt with the acquisition of mass spectral fragmentation rules for use by heuristic Dendral (Buchanan et al. 1976). It also dealt with the automatic formulation of theory for molecular fragmentation in mass spectrometry. Later, the same techniques were applied to the acquisition of ^{13}C NMR rules automatically (Schwenzer and Mitchell 1977). Thus, the Dendral group was attempting to provide help in the areas of knowledge acquisition for their expert system.

REFERENCES

J. Bachant and J. McDermott, "R1 Revisited: Four Years in the Trenches," *AI Mag.*, **5**(3), 21–32 (1984).

B. G. Buchanan, D. H. Smith, W. C. White, R. Gritter, E. A. Feigenbaum, J. Lederberg, and C. Djerassi, "Applications of Artificial Intelligence for Chemical Inference. XXII. Automatic Rule Formation in Mass Spectrometry by Means of the meta-DENDRAL Program," *J. Amer. Chem. Soc.*, **96**, 6168 (1976).

R. E. Dessy, "Expert Systems. Part I," *Anal. Chem.*, **56**, 1200A–1212A (1984).

R. O. Duda and E. H. Shortliffe, "Expert Systems Research," *Science*, **220**, 261–268 (1983).

F. Hayes-Roth, "Rule-Based Systems," *Commun. A.C.M.*, **28**, 921–932 (1985).

D. R. Hofstadter, *Gödel, Escher, Bach: An Eternal Golden Braid*, Vintage Books, New York, 1979.

J. Lederberg, "Topology of Molecules," in *The Mathematical Sciences: A Collection of Essays*, edited by the National Research Council Committee on Support of Research in the Mathematical Sciences (COSRIMS), MIT Press, Cambridge, MA, 1969.

R. K. Lindsay, B. G. Buchanan, E. A. Feigenbaum, and J. Lederberg, *Application of Artificial Intelligence for Organic Chemistry: The DENDRAL Project*, McGraw-Hill, New York, 1980.

R. A. Miller, H. E. Pople, Jr., and J. D. Myers, "INTERNIST-I, An Experimental Computer-Based Diagnostic Consultant for General Internal Medicine," *N. Engl. J. Med.* **307**, 468–476 (1982).

N. J. Nilsson, *Principles of Artificial Intelligence*, Tioga, Palo Alto, CA, 1980.

E. Rich, "The Gradual Expansion of Artificial Intelligence," *Computer*, **17**(5), 4–12 (1984).

G. M. Schwenzer and T. M. Mitchell, "Computer-Assisted Structure Elucidation Using Automatically Acquired C-13 NMR Rules," in *Computer-Assisted Structure Elucidation*, D. H. Smith (Ed.), American Chemical Society, Washington, DC, 1977.

H. A. Simon, *Sciences of the Artificial*, 2nd ed., MIT Press, Cambridge, MA, 1984.

A. M. Turing, "Computing Machinery and Intelligence," *Mind*, **59**(236), 433–460. (1950).

B. L. Webber and N. J. Nilsson, *Readings in Artificial Intelligence*, Tioga, Palo Alto, CA, 1981.

P. H. Winston, *Artificial Intelligence*, 2nd ed., Addison-Wesley, Reading, MA, 1984.

16

SPECTROSCOPIC LIBRARY SEARCHING AND STRUCTURE ELUCIDATION

The elucidation of the molecular structure of an unknown is a common task for chemists. A variety of approaches can be taken, and a variety of data can be used. Both physical and chemical methods are available and spectroscopic techniques are almost universally employed.

The process of elucidating unknown molecular structures involves the repeated application of three steps:

1. Posing of candidate structures.
2. Comparing them against available data.
3. Narrowing the list of candidate structures.

These steps can be called the posing, testing, and rejecting of hypotheses. Chemists have been performing these operations for nearly a century. The great increase in proficiency at this task is due to the tremendous expansion in data acquisition capabilities—largely in the spectroscopic area. The methods used in routine structure elucidation include infrared spectroscopy, visible-ultraviolet spectroscopy, NMR spectroscopy (proton and ^{13}C), and mass spectrometry.

16.1 MASS SPECTROMETRY

The variations of experimental design and data interpretation have led to major enhancements in the utility of mass spectrometry for structure elucidation. Ionization methods now include electron impact, chemical ionization, field ionization, fast atom bombardment, ion–molecule reactions in MS/MS instruments, and so on. High-resolution mass measurement allows assignment of empirical formulas for ions, which contains much valuable structural information.

One direct approach to structure elucidation involves searching through a file of reference spectra. The experimentally determined spectrum of the unknown compound is compared to each spectrum in the reference file. A list of the best matches is presented as the output. It is then up to the user to look at the structures corresponding to the best-matched spectra and to draw whatever conclusions are warranted.

Mass spectra are very well suited for computer-aided interpretation. Some reasons for this assertion are as follows. Mass spectral peaks occur at discrete values of mass, and for low-resolution mass spectra, they even occur at integral mass-to-charge ratios (except for metastable ion peaks and in some advanced mass spectrometry techniques). The mass spectra are rich in structural information since the ions observed are due to unimolecular decomposition reactions. Large numbers of reasonably good quality mass spectra have been gathered into collections, and with GC/MS instrumentation many spectra can be gathered per unit time. All modern mass spectrometers are computerized instruments, and the spectra are automatically collected and digitized, so saving the spectra is easy to do.

Library Searching of Mass Spectra

The purpose of searching through a library is to find those spectra in the reference file that most closely match the query spectrum. To perform a library search, one must have (1) a reference library of reasonable size, quality, and standardized format and (2) a searching algorithm that includes an effective method for assessing the degree of similarity between two spectra.

The method used to assess similarity between spectra depends on the details of how the spectra are stored, that is, their representation. With mass spectra, one representation is to use all the peaks. However, many studies have shown that abbreviated spectra can be used effectively. For example, one method for cutting the size of mass spectra is to store only the two most intense peaks in each 14-amu region (Hertz et al. 1971). The mass range of 14 was chosen because it is the mass of a methylene group.

PBM and STIRS

A pair of computerized mass spectral search systems for organic compounds have been developed by McLafferty and co-workers starting in the early 1970s. The searching system is named PBM (for probability-based matching), and the interpretative system is named STIRS (for self-training interpretive and retrieval system). A recent publication from this group (McLafferty et al. 1983) indicates that the file of mass spectra being used at that time was 79,525 spectra of 67,510 different compounds.

The key features of the PBM system that were incorporated into it from the beginning (McLafferty et al. 1974, Pesyna et al. 1976) are:

1. Weighting of the mass and abundance data.
2. Reverse searching.
3. A confidence measure for degree of match.

The mass and abundance data are weighted according to their statistical occurrence probabilities. The abundance values are weighted according to a lognormal distribution. The masses are given a value representing uniqueness that is based on their occurrence probability in the mass spectral data base empolyed.

The PBM system uses a reverse search strategy. To generate the file of reference spectra, the set of mass spectra on hand are analyzed to determine the probability of occurrence of each peak. The spectrum of each compound in the reference file is then represented by the 10 or 15 least probable peaks. This was justified both on the basis of information theory and also because it is in accordance with the way in which mass spectroscopists think. One consequence of this strategy is that it demands that the peaks of the reference spectrum be present in the unknown spectrum but *not* that all the peaks in the unknown spectrum be present in the reference spectrum. This is valuable in identifying components in a mixture.

The probability that a particular compound in the reference file is contained in the unknown is expressed by a quantity called the confidence index K (McLafferty et al. 1974). The K value for a particular reference compound is a summation over the individual values of K calculated for each peak being considered:

$$K = \sum K_j = \sum (U_j - A_j - D - W_j)$$

The four terms are probabilities. Here U_j is the uniqueness of the m/z value for the jth peak. It is a measure of the probability of a peak of each m/z-value to be prominent in a collection of representative mass spectra. The symbol A_j is a measure of the abundance of the jth peak in the reference spectrum. It acts as a modifier for the uniqueness probability in recognition of the fact that the magnitude of peaks as well as their location is important. The dilution factor D is a correction applied to the probability calculation because of the expected decrease in peak intensity due to dilution in a mixture. Values for U and A are based on pure compounds, and the dilution factor adjusts for this discrepancy. The value of D is set to zero for unknown spectra of pure compounds. Window tolerance W_j is related to how narrow the range of abundance values will be considered as a match.

The spectra that are found to be most similar during a search of the library file are reported along with a measure of the confidence attached to each retrieval. It is based on the statistical probability that the degree of match between the unknown and the reference spectrum occurred by coincidence.

The system uses negative information in the spectrum of the unknown because it takes into account peaks that are absent as well as those that are present. This is an important feature since the absence of characteristic mass spectral peaks can contain a great deal of structural information about the unknown.

McLafferty has proposed several statistical measures of the performance of PBM. Recall (RC) was defined as the fraction of relevant spectra actually retrieved:

$$RC = \frac{I_c}{P_c}$$

Reliability (RL) was defined as the fraction of retrieved spectra that is actually relevant:

$$RL = \frac{I_c}{I_c + I_f}$$

False prediction (FP) was defined as the proportion of spectra retrieved incorrectly:

$$FP = \frac{I_f}{P_f}$$

In these equations I_c is the number of correct retrievals, P_c is the possible number of correct retrievals, I_f is the number of false retrievals, P_f is the possible number of false retrievals.

Over the decade during which the PBM system has been under development, McLafferty and his co-workers have reported a number of improvements. One such improvement involved enhancing PBM's ability to deal with spectra of mixtures (Mun et al. 1981). The method involves subtracting the spectrum of the best-matched reference compound from that of the unknown mixture and then the resulting residual spectrum against the reference file a second time. It was found that this procedure was especially valuable for finding minor components in mixtures.

Another enhancement of the PBM system came from using an especially effective file-ordering method (Mun et al. 1981). The reference spectra are filed according to the mass of the peak that has the highest value of a function formed from the uniqueness and abundance of that mass in the data base. The uniqueness of the peak is the log value based on its occurrence in the data base, and the abundance is defined on a lognormal functional basis. During the searching procedure, the peaks in the unknown spectrum of highest $U + A$ value determine which of the indexes of the reference file are searched.

The STIRS Program

An alternative approach to the library searching implemented in PBM is to "interpret" the spectrum of the unknown. McLafferty and co-workers have developed a program to implement this idea, and it is called STIRS. STIRS has built into it a great deal of prior knowledge of mass spectrometry and mass spectral correlations, largely through the definition of 18 specific classes of mass spectral data. These classes were chosen because they are indicative of a variety of structural entities.

When STIRS is interpreting an unknown spectrum, it matches the data in each class for the unknown to the data in each class for each reference spectrum. It computes a match factor that is a quantitative measure of the degree of similarity of the unknown spectrum and each reference spectrum. For each class, the set of 15 best-matched reference spectra are saved for further analysis. The molecular structures of the compounds corresponding to each of the best-matched spectra are compared to seek common substructures. If a substructure is found to be present in many of the retrieved structures, it is probable that this substructure will also be present in the unknown as well.

STIRS can predict the molecular weight of the unknown compound too. The approach is based on the assumption that the best-matching compounds selected from the reference file will have primary neutral losses that are similar to those of the unknown compound. The program uses the mass values at the upper end of the unknown mass spectrum to compare to the upper mass values in the best-matched compounds. The predictions were reported to be better than 90% correct in one test of the system (Mun et al. 1981).

16.2 INFRARED SPECTRA

The infrared spectrum of a compound provides structural information that is complementary to that contained in the mass spectrum. It provides a method for "fingerprinting" organic compounds with regard to their functional groups. The presence of certain atom types, certain bond types, and functional groups can be confirmed by comparing the observed spectrum with correlation charts. The rules contained in the correlation charts have been built up over many years by experimentalists by empirical observations. Infrared spectra are commonly used in conjunction with other forms of spectroscopy for the elucidation of molecular structures of unknown compounds.

The most desirable way to use an infrared spectrum of an unknown compound is to compare it to the members of a file of identified spectra. A strong resemblance between the infrared spectrum of the unknown and one of the file spectra is strong evidence of structural similarity or even identity.

This searching option is only available if one has access to a large file of identified spectra of similar structural types to the unknown. Such collections of infrared spectra are costly to generate and maintain, so this means of interpreting the infrared spectrum of an unknown is often not feasible.

The alternative to library searching is interpretation of the spectrum by comparing to empirical correlation charts that relate infrared absorption bands to structural units. Quite extensive sets of empirical data are available in books. The automation of the reasoning process used by chemists to interpret infrared spectra has been attempted by several workers. Work started at Arizona State University (Woodruff and Munk 1977) and carried on at Merck Sharp and Dohme (Woodruff and Smith 1980, 1981, Tomellini et al. 1981, 1984a,b, Woodruff 1984) has yielded a computer program (called PAIRS for program for the analysis of infrared spectra). PAIRS is written in FORTRAN. The original version runs on an IBM mainframe computer and is available from the Quantum Chemistry Program Exchange as Program No. 426 (Woodruff and Smith 1981). A version has also been developed to run on a laboratory minicomputer, a Nicolet 1180 minicomputer. A more recent version, with substantial improvements, has been made available for VAX computers as QCPE Program No. 497 (Tomellini et al. 1985). PAIRS does not use or store a data base of infrared spectra, and it is therefore not subject to the limitations associated with all library search procedures. Woodruff (1984) stated: "PAIRS attempts to parallel as closely as possible the reasoning a spectroscopist uses in interpreting IR spectra.

PAIRS consists of several parts that interact with each other. The two most important parts of the system are (a) the set of empricial infrared rules that form a base of factual information for the program to use and (b) the methodology for manipulating the rules to obtain overall probabilities to report to the user. The rules are treated by the program as data, which is to say that they can be changed, updated, corrected, and so on, without the need to alter the program in any way. This way of organizing a program is used routinely in expert systems and has been purposely employed in PAIRS because of its great enhancement to program understanding and updating. The rules are input to the data base by the chemist using a specially designed language termed CONCISE (for computer-oriented notation concerning infrared spectral evaluation). The language was designed to make the input of the infrared rules as natural as possible to the spectroscopist.

When used for interpretation, PAIRS first requires the user to input the spectrum to be interpreted in digital format. The spectrum is digitized for the range 4000–5000 cm^{-1} with peak intensities coded from 1 (for very weak) to 10 (for very strong) and with peak widths of 1 (sharp), 2 (average), or 3 (broad). Ancillary information such as the sample state and the molecular formula are then input. The program then uses the rule set to perform the interpretation. Its output consists of probabilities for each functional group that is under consideration.

The design and construction of the rule set is the heart of PAIRS, and it is this aspect of the overall design of PAIRS that we will discuss in the following paragraphs.

The rules used by PAIRS are set up the same way that spectroscopists would do it, that is, by spectral regions that correspond to specific functional groups. Each functionality defines a class. The rules for each functionality are set up in the form of a hierarchical decision tree. Definite regions of the spectrum are examined sequentially. When an absorption is found in a region being examined, the conclusion reached is dictated by the way in which the treee is set up. The decision trees are related to the functionality. A specific example taken from Woodruff and Smith (1981) will clarify this method for representing infrared spectral information.

The example used by Woodruff and Smith was aldehydes. The information regarding absorptions that correlate with the presence of the aldehyde functionality is first found in standard reference books. The consensus of the rules for aldehydes is as follows: (a) a strong peak due to carbonyl appears between 1765 and 1660 cm^{-1}, and its exact position depends on the immediate surroundings of the carbonyl group; (b) two peaks of moderate intensity appear between 2900 and 2695 cm^{-1}, usually near 2820 and 2720 cm^{-1}. The 2720 cm^{-1} peak is usually sharp and resolved from interferences; it provides better evidence for aldehyde than the 2820 cm^{-1} peak.

These rules for aldehydes can be expressed as a decision tree, as shown in Figure 1. The entry point for this tree is at the left. The initial probability of aldehyde is zero. At each junction in the tree, a question is answered either yes or no, and the corresponding branch of the treee is followed. The first question to be answered is "Is there an absorption in the range 1765 to 1660 cm^{-1}?" If not, then the examination of the tree is finished, and PAIRS would continue to another section of its decision tree network (this is what "continue" means). If the peak is found, the Y branch is taken, and the region from 2750 to 2680 cm^{-1} is examined. If no peak is found, the aldehyde decision tree is abandoned. If a peak is found, the probability of aldehyde is set to 0.25. If the peak in this region is sharp, the probability of aldehyde is increased by 0.25. Then the region from 2850 to 2800 cm^{-1} is examined; if a peak is found in this region, the probability of aldehyde is increased by 0.25. Finally, the region from 1739 to 1681 cm^{-1} is examined, and if a strong peak is found, the probability of aldehyde is increased by 0.25. If all four tests were positive, the probability of aldehyde would be 1.0.

When these simple rules for aldehyde functionality were applied to real spectra, the results were not very satisfactory. However, these rules are quite simple and are used as an illustration of how the spectral information is incorporated into PAIRS. Woodruff and Smith (1981) present a much more complicated and complete decision tree for aldehydes that takes into account whether the sample was run in mineral oil, whether an acid functionality might be present, and some other factors. When this enhanced decision tree for aldehydes was tested, the results of the interpretations were much improved, as would be expected.

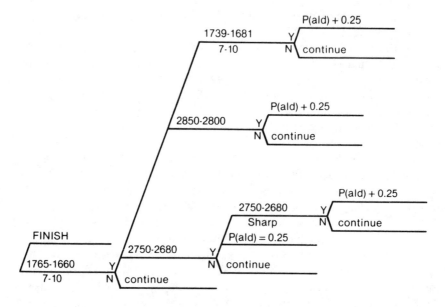

Figure 1. Decision tree expressing the interpretation rules for aldehyde compounds as used by PAIRS. (Reprinted with permission from *Progress in Industrial Microbiology*, Vol. 17, Elsevier Science Publishers, New York, 1983. Copyright 1983, Elsevier Science Publishers.)

The information present in the decision trees of PAIRS is entered by the chemist by using a special language called CONCISE. Smith and Woodruff (1984) have published a detailed description of CONCISE. It is designed to allow the chemist to generate, update, and maintain a set of infrared rules without having to know any computer language. Moreover, PAIRS was designed so that whenever the system makes faulty interpretations, the user can inspect the rules used. Then they can be modified, if desirable. At the very minimum, the user will be able to see exactly why PAIRS made a faulty interpretation.

In interpreting an infrared spectrum, PAIRS makes its way through all the decision trees available in its knowledge base. It then reports the probabilities of functional groups being present. In an early paper, Woodruff and Smith (1980) report that PAIRS had rules for more than 170 functional groups and functional group types. PAIRS considers organic compounds with atom types of C, H, O, N, S, F, Cl, and Br whose spectra have been taken in one of six sample states. A subsequent report on PAIRS described its modification to be able to deal with vapor-phase spectra (which require different rules than the original rules for condensed-phase spectra). PAIRS is able to deal with mixture samples just as easily as pure samples because no comparisons are done with reference spectra. Of course, if one of the components of a mixture is at a very low concentration, peaks would have lower than expected intensities, which could affect the results of the interpretation. The user must watch for effects of impurities on results.

PAIRS has been extensively tested with hundreds of infrared spectra of complex molecules, and the results show that the concept being investigated is valid. That is, one can develop a computerized assistant to help the chemist in a complicated task such as interpreting infrared spectra of organic compounds.

In more recent work, the PAIRS development group has reported the generation of programs that create infrared interpretation rules in CONCISE based on representative spectra (Tomellini et al. 1984a). The process is automated but also interactive. An automatic peak-finding routine provides information about peak positions, peak intensities, and peak widths. The program was developed using over 3300 vapor-phase infrared spectra. The rule generating capabilities of the program were tested by developing rules for saturated alcohols with a set of 51 vapor-phase spectra of 19 primary, 18 secondary, and 14 tertiary alcohols. Rules that could reliably differentiate between these three subclasses of alcohols were developed by the program. This work is an example of the automation of the knowledge acquisition part of an expert system.

REFERENCES

General

H. Abe, I. Fujiwara, T. Nishimura, T. Okuyama, T. Kida, and S. Sasaki, "Recent Advances in the Structure Elucidation System, CHEMICS," *Comput. Enhanc. Spectrosc.*, **1**, 55–62 (1983).

J. T. Clerc, E. Pretsch, and J. Seibl, *Structural Analysis of Organic Compounds by Combined Application of Spectroscopic Methods*, Elsevier, Amsterdam, 1981.

M. E. Munk, C. A. Shelley, H. B. Woodruff, and M. O. Trulson, "Computer-Assisted Structure Elucidation," *Fres. Z. Anal. Chem.*, **313**, 473–479 (1982).

S. Sasaki and Y. Kudo, "Structure Elucidation System Using Structural Information from Multisources: CHEMICS," *J. Chem. Inf. Comput. Sci.*, **25**, 252–257 (1985).

C. A. Shelley and M. E. Munk, "CASE: A Computer Model of the Structure Elucidation Process," *Anal. Chim. Acta*, **133**, 507–516 (1981).

D. H. Smith, *Computer-Assisted Structure Elucidation*, A.C.S. Symp. Ser., **54**, 1977.

Mass Spectra and PBM and STIRS

B. L. Atwater (Fell), R. Venkataraghaven, and F. W. McLafferty, "Matching of Mixture Mass Spectra by Subtraction of Reference Spectra," *Anal. Chem.*, **51**, 1945–1949 (1949).

H. E. Dayringer, G. M. Pesyna, R. Venkataraghavan, F. W. McLafferty, "Computer-Aided Interpretation of Mass Spectra. IX. Information on Substructural Probabilities from Self-Training Interpretive and Retrieval System (STIRS)," *Org. Mass Spectrom.*, **11**, 529–542 (1976).

K. S. Haraki, R. Venkataraghavan, and F. W. McLafferty, "Predication of Substructures from Unknown Mass Spectra by the Self-Training Interpretive and Retrieval System," *Anal. Chem.*, **53**, 386–392 (1981).

H. S. Hertz, R. A. Hites, and K. Biemann, "Identification of Mass Spectra by Computer-Searching a File of Known Spectra," *Anal. Chem.*, **43**, 681–691 (1971).

K-S. Kwok, R. Venkataraghavan, F. W. McLafferty, "Computer-Aided Interpretation of Mass Spectra. III. Self-Training Interpretative and Retrieval System," *J. Amer. Chem. Soc.*, **95**, 4185–4194 (1973).

S. R. Lowry, T. L. Isenhour, J. B. Justice, Jr., F. W. McLafferty, H. E. Dayringer, and R. Venkataraghavan, "Comparison of Various *K*-Nearest Neighbor Voting Schemes with the Self-Training Interpretative and Retrieval System for Identifying Molecular Structures," *Anal. Chem.*, **49**, 1720–1722 (1977).

D. P. Martinsen, "Survey of Computer Aided Methods for Mass Spectral Interpretation," *Appl. Spectrosc.*, **35**, 255–266. (1981).

F. W. McLafferty and D. B. Stauffer, "Retrieval and Interpretative Computer Programs for Mass Spectrometry," *J. Chem. Inf. Comput. Sci.*, **25**, 245–252 (1985).

F. W. McLafferty, R. H. Hertel, and R. D. Villwock, "Computer Identification of Mass Spectra. VI. Probability Based Matching of Mass Spectra. Rapid Identification of Specific Compounds in Mixtures," *Org. Mass Spectrom.*, **9**, 690–702 (1974).

F. W. McLafferty, S. Cheng, D. M. Dully, C-J. Guo, I. K. Mun, D. W. Peterson, S. O. Russo, D. A. Salvucci, J. W. Serum, W. Staedeli, and D. B. Stauffer, "Matching Mass Spectra Against a Large Data Base During GC/MS Analysis," *Int. J. Mass Spectrom. Ion Phys.*, **47**, 317–319.

I. K. Mun, D. R. Bartholomew, D. B. Stauffer, and F. W. McLafferty, "Weighted File Ordering for Fast Matching of Mass Spectra Against a Comprehensive Data Base," *Anal. Chem.*, **53**, 1938–1939.

G. M. Pesyna, R. Venkataraghavan, H. E. Dayringer, and F. W. McLafferty, "Probability Based Matching System Using a Large Collection of Reference Mass Spectra," *Anal. Chem.*, **48**,1362–1368 (1976).

R. Venkataraghavan, H. E. Dayringer, G. M. Pesyna, B. L. Atwater, I. K. Mun, M. M. Cone, and F. W. McLafferty, "Computer-Assisted Structure Identification of Unknown Mass Spectra," in *Computer-Assisted Structure Elucidation*, D. H. Smith (Ed.), American Chemical Society, Washington, DC, 1977.

Infrared Spectra and Pairs

G. M. Smith and H. B. Woodruff, "Development of a Computer Language and Compiler for Expressing the Rules of Infrared Spectral Interpretation," *J. Chem. Inf. Comput. Sci.*, **24**, 33–39 (1984).

S. A. Tomellini, and D. D. Saperstein, J. M. Stevenson, G. M. Smith, H. B. Woodruff, and P. F. Seelig, "Automated Interpretation of Infrared Spectra with an Instrument Based Minicomputer," *Anal. Chem.*, **53**, 2367–2369 (1981).

S. A. Tomellini, J. M. Stevenson, and H. B. Woodruff, "Rules for Computerized Interpretation of Vapor-Phase Infrared Spectra," *Anal. Chem.*, **56**, 67–70 (1984a).

S. A. Tomellini, R. A. Hartwick, J. M. Stevenson, and H. B. Woodruff, "Automated Rule Generation for the Program for the Analysis of Infrared Spectra (PAIRS)," *Anal. Chim. Acta*, **162**, 227–240 (1984b).

S. A. Tomellini, R. A. Hartwick, and H. B. Woodruff, "Automatic Tracing and Presentations of Interpretation Rules Used by PAIRS," *Appl. Spectrosc.*, **39**, 331–333 (1985).

S. A. Tomellini, G. M. Smith, and H. B. Woodruff, "PAIRS: Program for the Analysis of Infrared Spectra (VAX Version)," QCPE Program No. 497.

H. B. Woodruff, "Progress in Interpretation of Antibiotic Structures Using Computerized Infrared Techniques," in *Progress in Industrial Microbiology*, Vol. 17, M. E. Bushell (Ed.), Elsevier Scientific, Amsterdam, 1983, pp. 71–108.

H. B. Woodruff, "Using Computers to Interpret IR Spectra of Complex Molecules," *Trends Anal. Chem.*, **3**, 72–75 (1984).

H. B. Woodruff and M. E. Munk, "A Computerized Infrared Spectral Interpreter as a Tool in Structural Elucidation of Natural Products," *J. Org. Chem.*, **42**, 1761–1767 (1977).

H. B. Woodruff and G. M. Smith, "Computer Program for the Analysis of Infrared Spectra," *Anal. Chem.*, **52**, 2321–2327 (1980).

H. B. Woodruff and G. M. Smith, "Generating Rules for PAIRS: A Computerized Infrared Spectral Interpreter," *Anal. Chim. Acta*, **133**, 545–553 (1981).

H. B. Woodruff and G. M. Smith, "PAIRS: Program for the Analysis of Infrared Spectra," QCPE Program No. 426.

PART IV

GRAPHICS

17

GRAPHICAL DISPLAY OF DATA

Over the past 30 years there has been an explosion in the ability of science to generate experimental data because of increases in instrumental sophistication, computerization, and related events. In addition, the widespread availability of high-speed computers has made it possible to work on scientific problems of a magnitude not approachable previously. The amount of information that pours out of modern laboratory instrumentation or out of large computational simulations is simply enormous. This fact places a great burden on scientists in two ways: presentation of the results to themselves so that they can attempt to understand the meaning of the results and presentation of the results (often very much pruned and selected) to others so that information can be transmitted. One of the time-honored traditions of scientists is that of producing pictorial forms that we call graphs for presentation of information to ourselves and to others. Plots of variables are nearly always more informative than large tables of information. Humans can perceive patterns in plotted data with little effort, whereas making sense out of large sets of tabular material is nearly impossible. In many situations, a data set can be successfully analyzed by using only graphical methods. In other cases, graphical representations can enhance other forms of analysis.

The types of graphical representations that can be used to present data are completely dependent on the quantity and type of data to be presented. The simplest plots show how one dependent variable's value changes as a function of the value of the independent variable. However useful such simple plots may be, they are inadequate for many applications that are now common. They do not allow the presentation of complex information clearly. More sophisticated displays are needed.

The simplest plots show the values of two variables in a plane. It may be that the two variables are independent variables, or one might be a dependent variable. In either case, the eye of the viewer can comprehend a set of data points represented in this way as an entirety. Relationships that might be obscured within tabular presentation of the same values jump off the page when shown in a scatter plot.

The availability of computer hardware and software that supports graphics has led to the ability of scientists to generate plots at will, even extremely complicated ones. The excuses that kept us from plotting data by hand are no longer valid. The field of graphics has matured.

Display Devices

Terminals attached to digital computers that are used as graphical displays are cathode-ray tubes (CRTs). They generate images by directing electrons at a phosphorescent coating on the inside of a transparent glass tube. As computer graphics displays, they are essentially of two types: raster display and vector display. While both are based on the CRT, their characteristics are somewhat different and the types of graphics that they will support gracefully are somewhat different.

An ordinary television tube is the most common example of a raster device. It constructs pictures by scanning the face of the screen in repeated sweeps. The sweeping starts at the upper left of the screen and goes across one line before dropping an increment to sweep across the next line. By varying intensity (and color) within each picture element, a complete picture is generated. The entire screen is refreshed at a vary rapid rate, much faster than the human eye can perceive. Broadcast television in the United States operates with 525 scan lines; computer graphics terminals operate with anywhere from 256 to 1024 scan lines. The larger the number of scan lines, the higher the quality of the images.

Many graphics display terminals for computer use depend on this same technique. Therefore, they must render all output into a series of precisely placed dots on the screen. The degree of detail attainable on such a display is determined by the number of resolution elements on the screen, that is, how many potential dots there are on the screen and their spacing. Straight lines that are drawn on such a display parallel to either the x or y axis will have very good resolution, but straight lines drawn at angles can have noticeable steps in them. Figure 1 shows why this is so. The display must represent the straight line by the best series of discrete dots that is possible. If the resolution of the screen of the device is quite high, then the stepping may be barely perceptible. However, for the more common lower resolution screens, the stepping can be noticeable or even bothersome. However, it is an innate part of raster display devices.

Display of numerals or letters or other symbols on a raster display is done by filling in the appropriate small array of dots to form the character as needed. The quality of the representation depends on the size of the resolution element of the display device. The minimum array that can display the alphabet plus punctuation marks is a 5×7 grid. There are 35

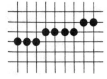

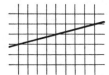

Figure 1. Raster and vector representations for straight lines.

intersections in such a grid, so any character can be represented by 35 bits indicating which intersections are brightened. Figure 2 shows a letter represented in the 5 × 7-bit array.

The second major type of graphics display device is the vector graphics type. Here, the electron beam that generates the display is switched on at a starting point and then moved from there directly to any other location on the screen, where the beam is switched off. Therefore, straight lines can be drawn accurately with no stepping at any angle at all. Figure 1 depicts the difference between raster and vector display of a straight line. The quantity of information that must be stored in order for the vector display to draw a line (the coordinates of the end points) is far smaller than for the raster display. Of course, display of numerals or letters on a vector display must be done differently than on the raster display. Each character must be drawn from a series of short, straight lines.

Display of Two-Dimensional Data

The most common form of graphical display is the simple *scatter plot*. The values of two variables are plotted in a rectangular array with each axis

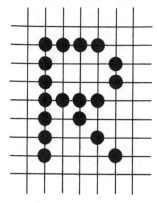

Figure 2. Alphabetic characters formed by the dot matrix method.

scaled to spread the data out over a large fraction of the available area. Scatter plots are the usual graphical depictions of the observed or dependent variable plotted against one independent variable. Such plots are commonly used to judge whether there is an apparent relationship linking the two variables plotted.

In addition to scaling of the two axes of scatter plots, one must decide on whether to make a functional transformation of one or both of the variables prior to plotting. For example, it is very common to use a logarithmic transformation on the variable plotted on the y axis to form what is commonly called a semilogarithmic plot. If there is an exponential relationship between the variable being plotted on the y-axis and the independent variable being plotted on the x-axis, as in an exponential decay curve, then the semilogarithmic plot makes the curve linear and therefore easier to perceive.

The most common output device is the printer. It may be a dot matrix printer or a line printer, but nearly all computers have some sort of hard copy output device. Printers can be used to generate low-resolution graphical output by placement of the print characters on the face of the paper. We used a simple printer plot to display the results of the kinetic simulations in Chapter 6. The subroutine PLOT implemented this simplified approach to printer plotting.

Display of Three-Dimensional Data

Many types of experiments generate data that is inherently three-dimensional. An example is tandem mass spectrometry. One type of tandem mass spectrometry is triple quadrupole mass spectrometry. The ions are formed by any of a number of ordinary ionization methods, and they are injected into the first quadrupole for mass analysis. Just one mass passes the first mass analyzer at any given time. The selected ion then passes into the second quadrupole, which is operated with only an RF field for focusing of the ions to prevent losses. In the second quadrupole, the ions undergo collisions with neutral gas molecules that cause simple fragmentation. The fragment ions enter the third quadrupole, which is operated as a mass analyzer, and only one ion at a time passes through it to the detector. The triple quadrupole experiment is done by scanning both the first and the third mass analyzers and recording intensities of ions as functions of the m/z settings on both analyzers. A complex three-dimensional data set is obtained. The data generated by the experiment puts in the experimenter's hands the formation and fragmentation modes of all the ions in the spectrum, which makes such a data set extremely valuable for elucidation of molecular structure.

Figure 3 shows such a data set in a three-dimensional representation as published in Yost and Enke (1981). The peaks are represented as narrow spikes because the resolution of the mass spectrometer is excellent and this

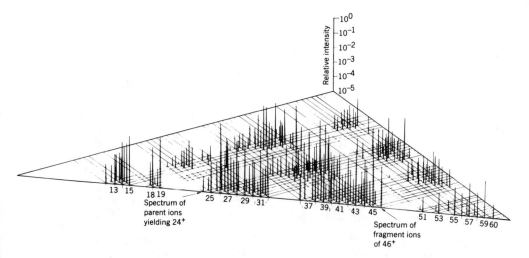

Figure 3. Three-dimensional fragmentation map from a triple quadrupole mass spectrometry experiment for isopropanol. [Reprinted from *American Laboratory*, **13**(6), 88–95 (1981). Copyright 1981 by International Scientific Communications, Inc.]

representation is a reasonably accurate picture of the appearance of the data. The molecule whose fragmentation pattern is shown is isopropanol. The front edge of the plot presents the ordinary mass spectrum of isopropanol, with the molecular ion peak at $m/z = 60$ on the right side. The plot shows the way to look at the data to see all the ions produced from fragmentation of the $m/z = 46$ ion. Also labeled is the line that shows all the ions that give rise to $m/z = 24$ ions. This type of information can be used to deduce the structure of molecules by comparing the data obtained to library spectra.

Another example of experimental data that is inherently three-dimensional comes from gas chromatography–mass spectrometry (GC/MS), where the observed data is mass spectral peak intensity as a function of time and the ratio of mass to charge. The complete experimental data can be plotted as time and mass-to-charge ratio along the two axes in the horizontal plane and ion intensity as a contour above the plane. Such a three-dimensional contour contains much more information than either the gas chromatogram or the mass spectrum alone.

Figure 4 shows an example plot of the results from a GC/MS experiment. The figure was generated by a software package called SuperIncos developed by Finnigan MAT. A 30-m GC column with SE-54 as the stationary phase was used along with temperature programming from 60 to 275 °C at a rate of 8 °C min^{-1}. The chromatogram appears along the horizontal axis, and directly above each GC peak appears the mass spectrum observed for that peak. The mass spectra are displayed for $m/z = 75$ to $m/z = 300$. A total of

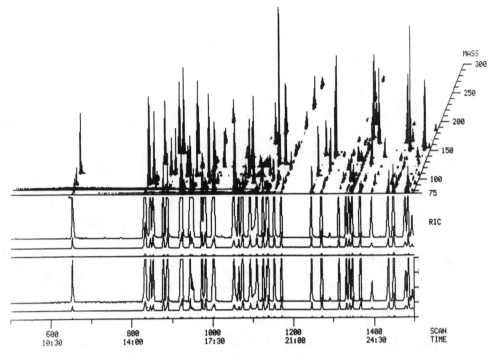

Figure 4. Three-dimensional representation of gas chromatography–mass spectrometry data. (Reprinted with permission from Finnigan Corporation.)

1500 mass spectra were scanned during the 26-min experiment. This display allows the researcher to view the entire set of experimental results at once so that inferences can be drawn on how to proceed.

The GC/MS plot differs significantly from the triple-quadrupole plot in that hidden lines are suppressed in order to make the display easier to understand. The plot is done so that the traces that are closer to the viewer obscure those that are more distant. This hidden line removal is very helpful in interpreting the contents of a complicated graphical display.

Representations of Surfaces and Hidden-Line Removal

Surfaces are often representative of the results of an experiment or a computation. Displaying of surfaces may be desirable in conjunction with such experiments or computations. The representation of a three-dimensional surface in two dimensions is the same problem that topographic map makers have wrestled with for a long time. The presentation of three-dimensional surfaces in the context of science is often handled in exactly the same way as in topographic maps, that is, by using contour lines. Looking down on a surface from directly above, along the normal, contours are drawn by tracing lines that represent the height of the surface above the surroundings. If this type of contoured surface is tipped on its side and if the hidden lines are removed, the resulting representation of a surface is readily understandable by viewing.

Figures 5 and 6 show two views of a surface that is represented by a grid. The surface being shown is the summation of three symmetrical Gaussian peaks with relative heights of 3, 4, and 5. The values of the function were calculated at each grid point, and the plot was constructed by joining these points with straight lines. Some of the lines have noticeable steps in them because the output was drawn on a dot matrix line printer. Figure 5 shows the relative heights very clearly, but Figure 6 is a better overall view of the relationship of the three peaks. These plots were generated with a modified version of the program PLOT3D (Watkins 1974).

The problem of hidden lines is one that has attracted a lot of research attention in computer science. It is easy to say how one wishes the graphical display to appear, but it is difficult to specify the operations in algorithmic terms. In addition, it is necessary that the algorithms be efficient, because identifying and removing hidden lines from displays is a computationally intensive task. We are discussing displays of surfaces represented by a grid of lines here, but when it is desired to display surfaces, the hidden surface problem becomes truly huge.

Display of Multivariate Data

When each observation or event is represented by many variables, display is not so straightforward as in the two-dimensional case. Multivariate data requires special methods for display.

A number of methods have been developed for the direct display of

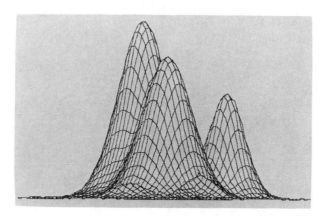

Figure 5. Plot of three Gaussian peaks showing the effect of hidden-line drawing on the representation of three-dimensional data.

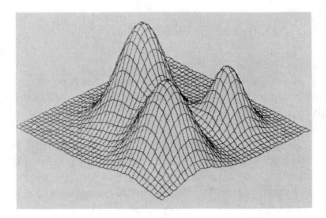

Figure 6. Plot of three Gaussian peaks showing the effect of hidden-line drawing and rotations on the representation of three-dimensional data.

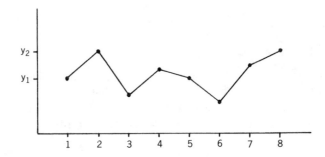

Figure 7. Linear and circular profiles for an eight-dimensional observation.

multivariate observations with no transformations. One of the simplest is the linear profile. The upper part of Figure 7 shows a linear profile constructed to represent an eight-dimensional observation. When a great many of such linear profiles are plotted next to each other for visual comparison, one can judge to some extent the similarities and differences among the observations.

A related method for the direct display of multivariate data is derived from the linear profile but with an additional touch. The ends of the linear profile are joined together to form a circular profile. The lower part of Figure 7 shows the circular profile representation of the same eight-dimensional observation. The circular profile was constructed by marking off eight equally spaced rays coming out of the center, which is marked with a cross. The ray pointing directly upward was arbitrarily assigned variable 1, and a point was placed on this ray at a distance y_1 away from the center. Then the next ray in a clockwise direction was marked with a point at a radial distance y_2 from the center. Similarly, the other six points were marked on their respective rays. Finally, the points were connected to form the circular profile. When many multivariate observations are being compared visually, the circular profiles are much easier to judge as similar or different than the linear profiles.

Figure 8 shows circular profiles for 20 of the five-dimensional data used for an example data set in the pattern recognition section. The top 10 patterns are from one class, and the lower 10 patterns are from the second class.

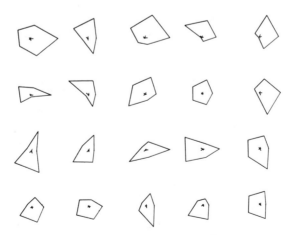

Figure 8. Circular profiles for 20 five-dimensional points, 10 from each of two classes.

REFERENCES

J. M. Chambers and B. Kleiner, "Graphical Techniques for Multivariate Data and for Clustering," in *Handbook of Statistics*, Vol. 2, P. R. Krishnaiah and L. N. Kanal (Eds.), North-Holland, New York, 1982.

J. M. Chambers, W. S. Cleveland, B. Kleiner, and P. A. Tukey, *Graphical Methods for Data Analysis*, Wadsworth International Group, Belmont, CA, 1983.

W. S. Cleveland and R. McGill, "Graphical Perception and Graphical Methods for Analyzing Scientific Data," *Science*, **229**, 828–833 (1985).

J. D. Foley and A. Van Dam, *Fundamentals of Interactive Computer Graphics*, Addison-Wesley, Reading, MA, 1982.

T. E. Graedel and R. McGill, "Graphical Presentation of Results from Scientific Computer Models," *Science*, **215**, 1191–1198.

S. L. Grotch, "Three-Dimensional Graphics for Scientific Data Display and Analysis," in *Chemometrics, Mathematics and Statistics in Chemistry*, B. R. Kowalski (Ed.), D. Reidel, Dordrecht, Netherlands, 1984, pp. 439–466.

D. L. Massart and L. Kaufman, *The Interpretation of Analytical Chemical Data by the Use of Cluster Analysis*, Wiley-Interscience, New York, 1983.

W. M. Newman and R. F. Sproull, *Principles of Interactive Computer Graphics*, 2nd ed., McGraw-Hill, New York, 1979.

C. F. Schmid, *Statistial Graphics. Design Principles and Practices*, Wiley-Interscience, New York, 1983.

E. R. Tufte, *The Visual Display of Quantitative Information*, Graphics, Cheshire, CT 1983.

P. H. Wang, *Graphical Representation of Multivariate Data*, Academic, New York, 1978.

S. L. Watkins, "Algorithm 483. Masked Three-Dimensional Plot Program with Rotations," *Commun. A.C.M.*, **17**, 520 (1974).

R. A. Yost and C. G. Enke, "An Added Dimension for Strucrture Elucidation Through Triple Quadrupole MS," *Amer. Lab.*, **13**, 88–95 (1981).

18

GRAPHICAL DISPLAY
OF MOLECULES

An important capability of computers is that of presenting complex information in pictorial form. In the context of chemistry in general and of molecular mechanics and the study of conformations in particular, this translates into the ability to display molecular structures. Presentation of molecules for visual analysis allows the chemist to view the model and judge its quality or seek insights based on the structure.

People are extremely accomplished at interpreting pictorial images because our primary information source is visual. Digital computers are extremely accomplished at performing routine computations tirelessly. Thus, graphical display of chemical information is a good example of man–machine cooperation. The machine does the repetitive work of managing the molecular model and displaying it, and the person views the result with higher level objectives in mind. The importance of molecular displays to modern chemistry is emphasized by the number of computer-generated views that appear in recent textbooks and journal articles and the recent startup of *Journal of Molecular Graphics* in the spring of 1983.

Viewing of objects on computer graphics display terminals in three dimensions is inherently more complex than the more common two-dimensional plots. The display screens are two-dimensional themselves. The solution to the mismatch between the two-dimensional screen and the three-dimensional molecular structure to be displayed on it is the use of projections. A projection is used to transform the three-dimensional molecule onto a two-dimensional projection plane. This projection plane can then be displayed directly.

A molecular structure consists of the identities of the atoms and the spatial coordinates of each atom. The coordinates either will be available in the Cartesian system or they can be transformed into it. Thus, each atom has an x, y, and z coordinate specified. To map such a three-dimensional object into 2-space for display, one alternative is to ignore the z coordinates of the atoms, that is, let them all be 0. This is equivalent to making what is called a *parallel projection*. A parallel projection is the view you get of a solid object when your eye is distant from the object, that is, the perspective is not apparent. In this instance, your line of sight would be parallel to the z axis. This is in distinction to perspective projections used in graphical display of objects in computer-aided design and engineering applications. For a complete discussion of these points, see a computer graphics text (e.g., Foley 1982).

The simplest displays of molecules are skeletal, or "stick," models. The bonds are represented by lines, and the atoms are understood to be present at the bond junctions. This is the same kind of graphical representation of molecules that is used in sketching on blackboards. This type of display can easily be programmed, but the picture generated can be confusing for all but quite simple structures. Figures 1a and 1b show a substituted norbornane molecule displayed in several graphical representations to allow comparison of information content.

Stick models can be made to appear more three-dimensional in several ways. Slow rotation of the stick model is feasible with some types of computer terminals, and this improves the clarity of the image. Such

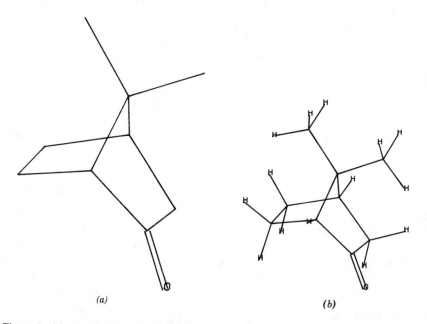

(a)   *(b)*

Figure 1. (a) Substituted nonbornane drawn in stick form with hydrogens suppressed. (b) Substituted norbornane drawn in stick form with hydrogens included. (c) Substituted nonbornane drawn with space-filling atoms and bonds. (d) Substituted norbornane drawn with space-filling atoms and intersections shown in CPK-like figure with hydrogens suppressed. (e) Substituted norbornane drawn with space-filling atoms and intersections shown in CPK-like figure with hydrogens included.

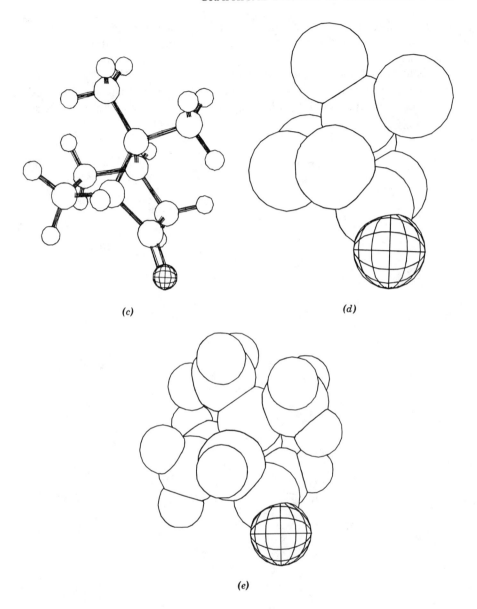

(c)

(d)

(e)

Figure 1. (*Continued*)

rotation gives the viewer a sense of the three-dimensionality of the image. Drawing of stereo pairs is also possible. Here, two copies of the same image are drawn side by side on the display terminal, differing from each other only in their degree of rotation about the vertical axis. Then the user can combine the two images either by using stereo glasses or defocusing the eyes. While these displays do demonstrate molecular architecture, they fail to convey information about the space-filling characteristics of the atoms. They provide little indication of how closely the nonbonded atoms approach each other.

A substantial increase in the information content of a drawing of a molecule is obtained when the atoms are represented by balls and the bonds by sticks. This type of molecular drawing is extremely common in X-ray crystallography, and most papers reporting research on crystallography

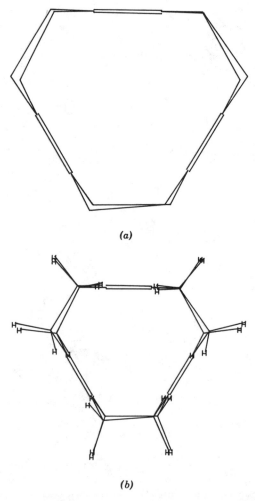

(a)

(b)

Figure 2. (a) Tetracyclooctadecatriene drawn in stick form with hydrogens suppressed. (b) Tetracyclooctadecatriene drawn in stick form with hydrogens included. (c) Tetracyclooctadeca-triene drawn with space-filling atoms in CPK-like representation with hydrogens suppressed. (d) Tetracyclooctadecatriene drawn with space-filling atoms in CPK-like representation with hydrogens included. (e) Tetracyclooctadecatriene drawn with space-filling atoms in CPK-like representation with hydrogens included and cross-hatching to suggest depth.

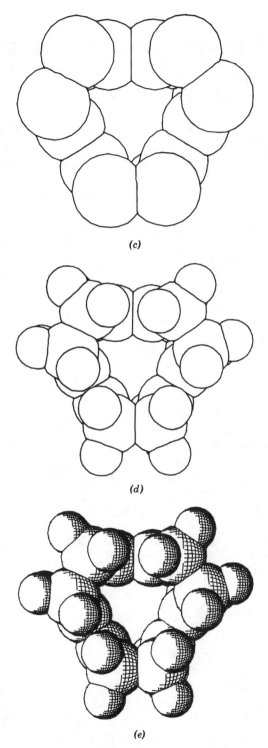

(c)

(d)

(e)

Figure 2. (*Continued*)

present graphical displays. Perhaps the best known display of this type is known as ORTEP (Oak Ridge thermal ellipsoid plot) (Johnson 1976). This display uses a graphics display terminal or hard copy device that will draw lines in one color. Figure 1c illustrates this type of representation.

An alternative approach for the display of molecules involves representing each atom by a sphere (e.g., Warme 1977, Smith and Gund 1978). Lines are drawn where atoms touch, and hidden lines are suppressed. This type of display shows a molecule as a three-dimensional object filling space. This type of display also uses lines only for the representation. Figures 1d and 1e show the substituted norbornane in this representation.

The visual display of molecular structures is a convenience for small, simple molecules. However, it becomes a virtual necessity as the molecular structures become more and more complex and more inherently three-dimensional. There is a class of compounds comprised of a number of six-membered rings jointed together by double bonds to form large overall rings. The second member of this class, with three six-membered rings, was reported in a paper by McMurry et al. (1984). It is a tetracyclooctadecatriene compound, and it is quite difficult to sketch on a blackboard. However, the use of computer graphics displays allows it to be displayed so that the overall structure is easily seen. Figures 2a–2e show this compound in several views as a stick figure and as a space-filling figure. The final figure, Figure 2e, shows how simple cross-hatching of the space-filling drawing can enhance the perception of three-dimensionality.

In addition to providing the user with visual images of molecules, computer-generated displays provide some other advantages. Any function that depends on three-dimensional model characteristics can be evaluated. For example, through-space distances between atoms that are not bonded to each other are readily available. Such distances could be computed as a function of conformational change. Secondly, two or more images can be superimposed on the screen. One of them could be rotated or otherwise manipulated while being compared to the other. A third example would be the increase in utility found if a molecular display system were linked to a data base of molecular structures, thereby allowing the user to browse through large sets of compounds seeking features of interest.

Display of Biomolecules and Macromolecules

Additional information can be included in a display when color is used. The advent of color display terminals and their incorporation into drug design and biomolecule research has led to an increase in the number of research articles that are accompanied by color space-filling displays (e.g., Knowlton and Cherry 1977, Max 1979, Lesk and Hardman 1982, Feldmann 1983, Langridge et al. 1981, Bash et al. 1983). Max has implemented a routine that simulates the surface of molecules to give a visual impression of the overall surface of a structure. Langridge and co-workers have developed a system to represent molecular surfaces with dots of color, and they have used this display to study interactions between biomolecules. An example of the publication of a color display in a research article whose focus is on structure–activity relations is given in Hansch et al. (1982).

REFERENCES

P. A. Bash, N. Pattabiraman, C. Huang, T. E. Ferrin, and R. Langridge, "Van der Waals Surfaces in Molecular Modelling: Implementation with Real-Time Computer Graphics," *Science*, **222**, 1325–1327 (1983).

R. J. Feldmann, "The Design of Computer Systems for Molecular Modeling," *Ann. Rev. Biophys. Bioeng.*, **5**, 447–510 (1976).

R. J. Feldmann, "Directions in Macromolecular Structure Representation and Display," in *Computer Applications in Chemistry*, S. R. Heller and R. Potenzone (Eds.), Elsevier, Amsterdam, 1983.

J. D. Foley, *Fundamentals of Interactive Computer Graphics*, Addison-Wesley, Reading, MA, 1982.

C. Hansch, R. L. Li, J. M. Blaney, and R. Langridge, "Comparisons of the Inhibition of *Escherichia coli* and *Lactobacillus casei* Dihydrofolate Reductase by 2,4-Diamino-5-(Substituted-benzyl) Pyrimidines: Quantitative and Structure–Activity Relationships, X-Ray Crystallography, and Computer Graphics in Structure–Activity Analysis," *J. Med. Chem.*, **25**, 777–784 (1982).

C. H. Hassall, "Computer Graphics as an Aid to Drug Design," *Chem. Brit.*, **21**, 39–46 (1985).

A. J. Hopfinger, "Computational Chemistry, Molecular Graphics and Drug Design," *Pharm. Int.*, **5**, 224–228 (1984).

C. Humblet and G. R. Marshall, "Three-Dimensional Computer Modeling as an Aid to Drug Design," *Drug Devel. Res.* **1**, 409–434 (1981).

C. K. Johnson, "ORTEP-II: A Fortran Thermal-Ellipsoid Plot Program for Crystal Structure Illustrations," ORNL-5138, Oak Ridge National Laboratory, March 1976.

K. Knowlton and L. Cherry, "ATOMS: A Three-D Opaque Molecule System—for Color Pictures of Space-Filling or Ball-and-Stick Molecules," *Comput. Chem.*, **1**, 161 (1977).

P. Kollman, "Theory of Complex Molecular Interactions: Computer Graphics, Distance Geometry, Molecular Mechanics, and Quantum Mechanics," *Acc. Chem. Res.*, **18**, 105–111 (1985).

R. Langridge, T. E. Ferrin, I. D. Kuntz, and M. L. Connolly, "Real-Time Color Graphics in Studies of Molecular Interactions," *Science*, **211**, 661–666 (1981).

A. M. Lesk and K. D. Hardman, "Computer-Generated Schematic Diagrams of Protein Structures," *Science*, **216**, 539–540 (1982).

J. E. McMurry, G. J. Haley, J. R. Malt, J. C. Clardy, G. Van Duyne, R. Gleiter, and W. Schafer, "Tetracyclo (8.2.2.2^{2,5}.2^{6,9})octadeca-1,5,9-triene," *J. Amer. Chem. Soc.*, **106**, 5018–5019 (1984).

N. L. Max, "ATOMLLL: A Three-D Opaque Molecule System (Lawrence Livermore Laboratory Version)," UCRL-52645, Lawrence Livermore National Laboratory, January 1979.

S. Ramdas and J. M. Thomas, "Computer Graphics in Solid-State and Surface Chemistry," *Chem. Brit.*, **21**, 49–52 (1985).

G. M. Smith and P. Gund, "Computer-Generated Space-Filling Molecular Models," *J. Chem. Inf. Comp. Sci.*, **18**, 207 (1978).

J. G. Vinter, "Molecular Graphics for the Medicinal Chemist," *Chem. Brit.*, **21**, 32–38 (1985).

P. K. Warme, "Space-Filling Molecular Models Constructed by Computer," *Comput. Biomed. Res.*, **10**, 75–82 (1977).

INDEX

Accuracy, 19
Adjacency matrix, 157
Adsorption, 109
Algebraic laws, 27
Algorithm, 9
ALPATH, 165
Analysis of residuals, 38
Argon atom, 101
Artificial intelligence, 212
ASCII, 4
Assembly, 16
Atom-by-atom search, 176
Atom path code, 164
Autoscaling, 192

Bayesian discriminant, 196
Benzene, 108
Buckingham potential, 183

CAS, 8
CAS ONLINE, 175
Central Limit Theorem, 97
Chemical structure information handling, 143
Chemisorbed species, 107
Chi-square, 97
Circular profile, 241
Classification, 195
Clustering, 200
Compilation, 16
Confidence index, 221
CONGEN, 217
Connection table, 150
Contour lines, 238
Convergence criteria, 130
Correlation coefficient, 38
    multiple, 55
Coulomb potential, 99

Cramer's rule, 87
CTIS, 153
CTPR, 153
CTTEST, 153
Curve fitting, 33
    transformation, 42

DATARD, 58
Debugging, 12
Decision surface, 195
Decision tree, 225
DENDRAL, 162, 216
Determinants, 87
Differential equations, 72
    forward-tracing, 72
    predictor-corrector, 83
    Runge–Kutta, 73
Discrete event simulation, 109
Discriminants, 195
    nonparametric, 197
    parametric, 197
Display, 193
Display devices, 234
DLIN, 92

Eadie–Hofstee, 49
EBCDIC, 4
Editing, 16
Enzyme kinetics, 50
Enzyme reactions, 47
Epsilon, machine, 28
ERROR, 137
Error, 19
    addition and subtraction, 24
    averaging, 25
    multiplication and division, 24
    propagation, 22

Execution, 16
Expert systems, 212
    examples, 216
EXPFIT, 44

Factor space, 126
Factorial, 67
F distribution, 56
FINISH, 60
Floating point number system, 26
Force field, 179
Fourier transform, 7
Fragmentation map, 237
Fragment codes, 144

Gamma function, 67
Gaussian quadrature, 69
Gauss–Jordan elimination, 89
Gauss–Legendre integration, 70
Gauss–Seidel, 88
GJE, 91
Graphical display, 233
Graphics, 9
Graph theory, 156
Grid searches, 126

Hamilton's equations, 105
Heuristic, 215
Hidden lines, 239
Hierarchical clustering, 201
Hooke's law, 181
HPLC simulation, 109

If-then rules, 215
IMSL, 28
Infrared spectra, 223
Integer, 4
Integration, 62
INTR, 119
Ionization, 108
Ion trajectory, 113
Isomer enumeration, 162
Iterative, 96
Iterative methods, 31

Karhunen–Loeve transform, 193
K nearest neighbor, 197
Knowledge representation, 213

Laws of algebra, 27
Least squares, 33
    examples, 39
    exponential function, 43
    linear, 33
    nonlinear, 133
    polynomial, 39

    weighted, 36
Lennard–Jones potential, 99, 183
Library searching, 219
Linear equations, 85
Linear learning machines, 198
Linear notation, 144
Linear profile, 241
LINEG, 92
Lineweaver–Burk, 48
LLS, 44, 51
LM, 201

Machine epsilon, 28
Mapping, 193
Mass spectra, 219
Mass spectrometry, 110
Mathieu equation, 112
Matrix methods, 85
Mean, 20
MERCURY, 101
Methane molecule, 101
Michaelis-Menten equation, 48
    example, 49
MICMEN, 50
Mixtures, 208
MLR, 53
Modeling, 201
Module, 11
Molar refraction, 23
Molecular dynamics, 98
Molecular graphics, 243
Molecular mechanics, 179
Molecular path code, 163
Molecular similarity, 165
Monte Carlo, 96
Morgan algorithm, 159
Morse function, 102
MREG, 58
Multidimensional problems, 126
Multiple correlation coefficient, 55
Multivariate data, 239

Newton's equation, 104
Newton–Raphson, 184
Ni (001) surface, 108
NISR, 66
NMR magnetic fields, 131
Nonbonded potential, 183
Nonlinear mapping, 194
Nonnumerical methods, 141
Nonparametric discriminant, 199
Normalization, 192
Number systems, 5

Optimization, 125
    applications, 131

Optimization of code, 14
ORTEP, 248

PAIRS, 224
PATH, 165
Path counts, 162
Pattern recognition, 186
    applications, 207
Pattern vector, 187
PBM, 220
PLOT, 79, 120
PLOT3D, 239
PLTEND, 121
Potential energy, 101
Potential energy surface, 100
Precision, 20
PRED, 205
Preprocessing, 192
Principal components analysis, 193
Probability density function, 21
Production rules, 215
Programming, 11
Pseudorandom numbers, 96

QCPE, 224
QMAS, 112
QMF, 110
QUAD, 118
Quadrupole mass analyzer, 110
Quantum Chemistry Program Exchange, 180

RAND, 96
Random error, 19
Random numbers, 96
Raster display, 234
Reaction dynamics, 100
Real, 4
Reflection vertex, 128
Registration, 158
Regression linear, 53
    multiple, 53
Residuals, 38
Response surface, 127, 134
RK4, 74
Runge–Kutta, 73
RUNKUT, 76

Scaling, 192
Scattering, 104
Scatter plot, 235
Scientific notation, 20, 26
Screens, 173
Secondary ion mass spectrometry, 104

Significant figures, 20
SIMCA, 201
Similarity, 187
Similarity of molecules, 165
Simplex, 125, 133
    applications, 131
SIMPLX, 133
SIMS, 104
Simpson's rule, 65
Simulation, 96
    chromatography, 108
    discrete event, 109
    water, 98
Single factor variation, 126
Single-precision, 4
SSE, 34, 55
Standard deviation, 20
Statistics, 19
Stationary phase, 109
Steepest descent, 184
Stereo pairs, 245
Stick models, 244
STIRS, 220
Strain energy, 180
Structure-activity, 208
Structure elucidation, 219
Subgraph isomorphism, 172
Substructure searching, 171
Sum-squared error, 34, 55
Surface representation, 238
Surface scattering, 104
SWEEP, 59
SWEEP operator, 57
Syntax errors, 12
Systematic errors, 19

Taylor series, 65, 73
Topological representation, 144
Torsional angle, 182
TRAIN, 204
Trajectory calculation, 100
TRAJEK, 118
TRANS, 45

Unambiguousness, 144
Uniqueness, 144

Variable selection, 56
Vector display, 234
Vertex, 126

Weight vector, 195
Wiswesser line notation, 145